Nigel Henbest and Heather Couper take us on a personal tour of the Milky Way in this delightful guide to our home Galaxy. This book is written in non-technical language, for the general reader and the amateur astronomer. The lucid text is supported by a dazzling portfolio of colour photographs of stars, star clusters and gas clouds in the Milky Way. Specially created maps locate the hundreds of tourist spots visited in this journey through space.

The guide introduces the nature and structure of our Galaxy as a whole. All types of objects that it contains are then described. These include mysterious black holes, the giant globular clusters of stars, every kind of star, regions of star formation, exploding stars and their remnants and the enigmatic objects at the galactic centre, as well as the giant halo surrounding the Galaxy.

THE GUIDE TO THE GALAXY

THE GUIDE TO THE GALAXY

Nigel Henbest and Heather Couper

Picture editor: Michael Marten

CAMBRIDGE
UNIVERSITY PRESS

Published by the Press Syndicate of the University of Cambridge
The Pitt Building, Trumpington Street, Cambridge CB2 1RP
40 West 20th Street, New York, NY 10011-4211, USA
10 Stamford Road, Oakleigh, Melbourne 3166, Australia

First published 1994

Photoset in ITC Galliard by Rowland Phototypesetting Ltd,
Bury St Edmunds, Suffolk
Printed and bound in Great Britain by Butler & Tanner Ltd,
Frome and London

A catalogue record for this book is available from the British Library

Library of Congress cataloguing in publication data

Henbest, Nigel.
The guide to the galaxy/Nigel Henbest and Heather Couper.
p.   cm.
Includes index.
ISBN 0-521-30622-1. – ISBN 0-521-45882-X (pbk.)
1. Milky Way.   I. Couper, Heather.   II. Title.
QB857.7.H46   1994
523.1'13-dc20   93-8859   CIP

ISBN 0 521 30622 1 hardback
ISBN 0 521 45882 X paperback

RO

CONTENTS

ACKNOWLEDGEMENTS

This book would not have been possible without the many individuals who located, supplied or processed imagery or information especially for us. Particular thanks for their time and trouble to: Marion Blonk, Mary Chibnall, Dennis Di Cicco, Coral Cooksley, Thomas Dame, Gary Evans, Martha L. Hazen, Peter Hingley, Gaylin Laughlin, R. A. Marriott, Dorothy Schaumberg, Jean-Pierre Sivan, John Wells, Farhad Yusef-Zadeh.

The picture research was by Caroline Erskine, and the Galaxy maps by Julian Baum.

Our thanks also to John Storey, for permission to reproduce part of his paper "The Detection of Shocked CO Emission From G333.6-0.2" (first published in the *Proceedings of the Astronomical Society of Australia*, vol. **5**, p. 566, 1984).

CHAPTER 1

The discovery of our Galaxy

Over the past few years, dozens of books – ranging from the academically abstruse to the unapologetically popular – have been written about our Solar System. The reasons aren't hard to find. In less than a couple of generations, spaceprobes have completely revolutionized our knowledge of the Sun and planets – to an extent summed up by the comment that 'pretty well all we thought we knew was nonsense'.

Until the early 1970s, the planets were just little dots in the sky that moved from night to night. Blurred photographs taken from below the Earth's shifting sea of atmosphere revealed only vague, indistinct markings. Then came a flotilla of spaceprobes with stirring names like Venera, Mariner, Viking, Pioneer and Voyager. Almost overnight, they trans-formed the remote dots and their encircling moons into *real* 'new worlds' – like the continents discovered by the Renaissance navigators – with mountains and valleys, belching volcanoes, icy poles and dark, mysterious canyons. Our world is just one of this huge family of highly individual worlds.

But we are also part of a much bigger family. The Sun – our local star – is one of an estimated 200 billion making up our 'star-city', the Galaxy. For the past 60 years, we have known that our Galaxy is a spiral wheel of stars, with the Sun located somewhere out near the rim. The Galaxy is so vast, however, that we cannot – at present – explore it with space probes in the way we are surveying our Solar System.

And yet our knowledge of the Galaxy, too, has been undergoing a revolution – albeit quieter than the one that hit the Solar System. Living as we do inside the Galaxy, it has always been difficult to assess its structure: the cosmic equivalent of 'not seeing the wood for the trees'. But now we have the means to penetrate the undergrowth. Thanks to our ability to place instruments above the Earth's atmosphere, and to innovations in electronics, we can peer out through new windows onto the Universe. The light we see from the stars gives us only one view. Radio waves, millimetre waves, infrared radiation, ultraviolet, X-rays and gamma rays show us completely different pictures of the Universe – and of our Galaxy.

The new views have given us a fresh understanding of the Galaxy. No longer is it just an assemblage of stars moving sedately through empty space, affecting each other only by the long-range bonds of gravity. We now see the Galaxy as an active, and interactive, place. Space between the stars is far from empty: indeed, we can now map the swirls and clouds in the interstellar gas as we can in the Earth's atmosphere. In the densest clouds, we can now detect stars being born. And we can measure how dying stars churn up the interstellar gas, and promote the birth of new stars. Just as some scientists have said that we should see the Earth's land, sea and living beings as a complete entity, 'Gaia', so we now see the Galaxy as more than just the sum of its parts.

This book is a celebration of the first clear, pan-spectral view of our own star-city, and the new 'holistic' interpretation of the Galaxy. It is a book that could not have been written until now. The maps, in particular, could not have been drawn up even a few years ago. At last we are beginning to see our Galaxy as a whole.

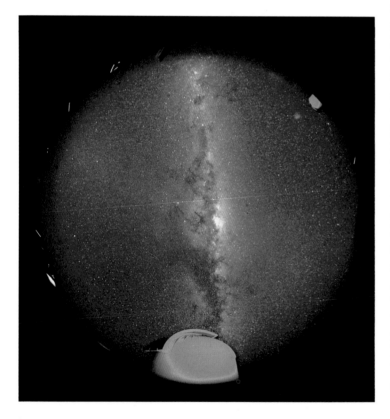

This dramatic 'all-sky' photo of the southern heavens – taken at the Las Campañas Observatory in Chile – shows the Milky Way arching overhead. At the centre, you can easily see the 'galactic bulge' that surrounds the nucleus of our Galaxy. This image captures the true nature of the Milky Way: as a spiral galaxy seen edge-on from a vantage point outside.

Origin of the Milky Way: milk gushes from the breast of the goddess Juno as she nurses the boisterous infant Hercules. But because the milk spurted into the sky – and not into the baby's mouth – Hercules missed out on his chance of immortality. This Greek myth is typical of many cultures' legends about the Milky Way, who also saw it as a river winding its way amongst the stars.

It has taken us centuries to reach this stage. This is because our view from the inside looking out reveals little: all we can see of our Galaxy is a misty band of light threading its way across our skies. In the clear, dark nights of antiquity, this ghostly band was well known, and provided the stuff of many legends. To the Romans and Greeks, it was, respectively, the 'Milky Way' (*Via Lactea*) and the 'Milky Circle' (*Kiklos Galaxias*). The Greeks even came up with the first theory for the origin of the Galaxy: that the Milky Way was a stream of milk which gushed from the breast of the goddess Juno as she nursed the thirsty infant Hercules.

Many civilizations saw the Milky Way as a path or river winding its way through the heavens. While the North American Indians regarded it as the route of ghosts on their way to the 'land of the hereafter', the Eskimos saw it as a path of glowing ashes to guide travellers safely back home. The Australian Aborigines had a different interpretation. Under desert skies so densely packed with stars, they saw patterns in the sky where *fewer* stars were visible. In the darker areas strung out along the length of the Milky Way, they were able to make out the shape of a giant emu spanning their skies.

Until January 1610, all our interpretations of the nature of the Milky Way were pure speculation. Then Galileo Galilei turned his 'optik tube' – the first-ever astronomical telescope – to the skies above Renaissance Padua. 'I have observed the nature and material of the Milky Way', he wrote. 'With the aid of a telescope, this has been scrutinised so directly, and with such ocular certainty, that all the disputes which have vexed philosophers through so many ages have been resolved, and we are at last free from wordy debates about it. The Galaxy is in fact nothing but congeries of innumerable stars grouped together in clusters. Upon whatever part of it the telescope is directed, a vast crowd of stars is immediately presented to view, many of them rather large and quite bright, while the number of smaller ones is quite beyond calculation'.

Unfortunately, few astronomers followed up Galileo's discovery. Galileo himself died in 1642, and in that year was born a man who would keep astronomers busy for the next century-and-a-half: Isaac Newton. He is chiefly remembered, of course, for his theory of universal gravitation, which describes how bodies interact with one another through the force of gravity. Proving Newton right (or wrong) became a major obsession with the astronomical community. At the same time, telescopes were undergoing vast improvements – Newton himself had made his own contribution here with the design of the 'Newtonian' reflecting telescope – and they had begun to reveal an astonishing wealth of detail on the Moon

The first person to observe the Milky Way through a telescope, Galileo discovered that it was made of 'innumerable stars'. But unfortunately, astronomers did not really pick up on his important findings until nearly two centuries later.

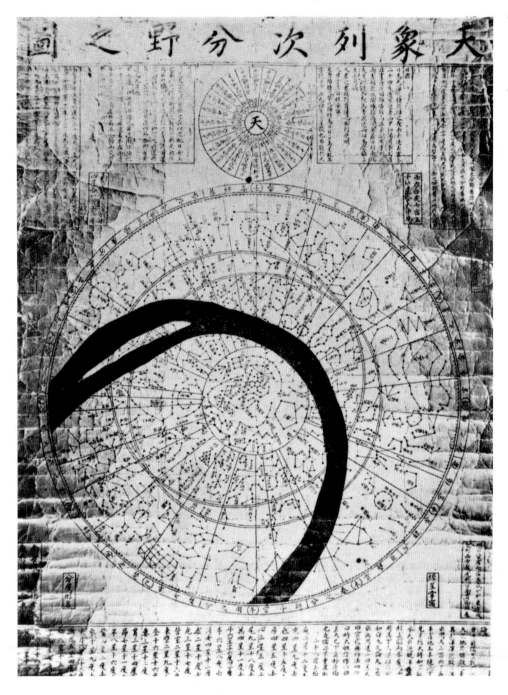

This planisphere (star map) of 1395 shows the stars visible from the latitude of Korea. It is based on an earlier map of AD 672, and it reveals how the oriental astronomers divided up the sky into many very small constellation patterns. The Milky Way is represented by a black band, and looks very much like a river crossing the sky.

FACING PAGE In this eighteenth century map of the southern sky, the Milky Way – marked Via Lactea – threads its way through some old and some new constellations. The great ship Argo and Centaurus (the centaur) date back to antiquity, but Musca (the fly) and Chameleon are more recent. Robur Carolinum (Charles's oak), introduced by Edmond Halley to flatter King Charles II in 1678, has long since dropped out of use: it lay in the region of the great gas cloud now known as the Carina Nebula.

and planets. All in all, there was plenty for astronomers to do close to home.

But the philosophers did not desert the Galaxy. In fact, the first person to attempt to delineate the Galaxy as we know it today was a philosopher called Thomas Wright. Wright was born in 1711 in Durham – into a rather lowly family – but he gained support in high places by teaching young noble ladies. He became something of a gentleman himself, and certainly a great thinker in matters of 'natural philosophy'. But his over-riding commitment was to God; he was always looking for confirmation that God existed in abundance.

To Thomas Wright, the shape of the Milky Way was a classic example of God's design. He saw it as perhaps a slab of stars, in the centre of which was a source of supernatural energy from which flowed goodness, morality and wisdom in abundance. Or it might instead be a sphere of stars – one amongst many other spheres of stars in the Universe.

Wright published his remarkably 'modern' ideas as *An Original Theory, or New Hypothesis of the Universe*. Unfortunately, his ideas were not taken terribly seriously by his contemporaries, chiefly because Wright was inconsistent. He rejected his 'original theory' a little later on, and replaced it with the notion that stars were giant volcanoes belching in the darkness.

Meanwhile, a young Prussian philosopher called Immanuel Kant had read a garbled account of Wright's theory in a German newspaper. Kant was a very different kind of philosopher from Wright. He was a trained mathematician; and he wanted to see if he could put Wright's elegant ideas onto a scientific footing.

Kant reasoned from the principle of uniformity. He argued in a logical and beautiful manner. If the Milky Way is a flat distribution of stars, then – surely – some of the other fuzzy, nebulous patches visible in the sky are 'Milky Ways', too. Our Galaxy, then, is just one of many in an enormous Universe.

Kant wrote up his findings in a small pamphlet entitled *General Natural History and Celestial Theory, or Research into the Constitution and Mechanical Origins of the Whole World Structure based on Newton's Law*. It did not take the Earth by storm. First, it was published anonymously. Then the publisher went bankrupt. Finally, most of the leaflets ended their days mouldering away in a warehouse. Immanuel Kant today is chiefly remembered for his work on human reasoning – and not for his investigations into the structure of the Universe.

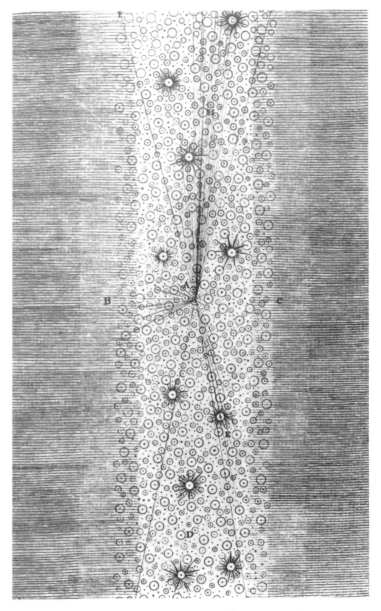

Eighteenth-century philosopher Thomas Wright believed the Milky Way to be a slab of stars with a source of 'supernatural energy' issuing from the centre. The description sounds rather like our modern conception of the Galaxy, with energy emerging from the disturbed galactic nucleus. But Wright's model was based only on his imagination, and was never taken seriously.

was fascinated by anything scientific. George appointed Herschel his personal astronomer, and set him up in a house close to the Royal Court at Windsor. Here Herschel's only duties were to show the wonders of the heavens to the Royal Family and visiting foreign dignitaries from time to time. Otherwise, he was able to indulge his passion for building bigger and bigger telescopes.

His labours eventually culminated in what was then the biggest telescope in the world: a monster with a light-collecting mirror 1.2 metres (4 feet) across, known usually as the '40-foot' after the length of its tube. The 40-foot and its surrounding merry-go-round of a mounting were, unfortunately, very unsafe to use. While the observer peered into the eyepiece from some high position atop the structure, engineers would move the telescope to different areas of the sky by means of a complicated system of ropes and winches. Someone once compared observing with the instrument to the act of shaving with a guillotine – so it was hardly surprising that the King's workmen downed tools and refused to work with it again!

Herschel's motivation for building ever-larger telescopes was not to create a technical challenge for himself. Instead, he needed to see further than anyone before him in order to fulfil his ambition: to make a really deep survey of the heavens. Herschel's greatest desire was to discover how the stars were really distributed through space.

To study the whole sky would have taken many lifetimes, so Herschel – using his smaller '20-foot' telescope – decided to sample 700 different regions instead. In each region, he counted the number of stars of different brightnesses. Herschel had no way of knowing the distances to any of these stars (the first stellar distance was not to be measured until 1838), so he made a sensible guess. He assumed to start with that all stars are the same intrinsic brightness – the same as Sirius, the brightest star in the sky. It followed then that the brighter stars in a given area were nearer, and the fainter stars further away. And so, in this way, Herschel was able to work out a 'scale model' of the star distribution – the first real map of our Galaxy.

Herschel's map revealed a flat distribution of stars, with the Sun close to the centre. It was nicknamed the 'grindstone model', although its edges were much more ragged than any terrestrial grindstone. These ragged edges were a result of dark rifts that Herschel could see in the Milky Way – the same

William Herschel (1738–1822) is remembered as the first person in history to discover a planet – Uranus. But his discovery of Uranus was just part of a very ambitious survey of the whole heavens, in an attempt to discern the structure of the Universe.

All these blind alleyways meant one thing: by the mid-eighteenth century, there was very little knowledge of, or interest in, the Milky Way. Then, in 1781, a musician and amateur astronomer named William Herschel did something that no man before had done. He discovered a new world at the frontiers of our Solar System – the planet Uranus.

Herschel's discovery endeared him to King George III, who

BOX 1. **The father of stellar astronomy**

William Herschel is chiefly remembered as the first man in history to discover a new planet. A musician and composer by profession, but a dedicated amateur astronomer by inclination, Herschel became a highly skilled telescope-maker. With his telescopes, which were better than most that were in the hands of professionals at the time, Herschel hoped to achieve the impossible: to measure the distance to a star.

To do this required a thorough survey of the sky, so Herschel spent every clear night in his back garden in Bath scrutinizing the stars. On 13 March 1781, he found an object that he noted in his observing log as 'curious'. It showed a disc; it was greenish in colour; and over a period of days, it moved against the background of stars. Herschel supposed that it was a comet. But its orbit revealed otherwise. This was an undiscovered planet, twice as far away from the Sun as Saturn. By finding this new world – later named Uranus – Herschel doubled the size of the Solar System at a stroke.

As a reward for his discovery, Herschel was appointed a Royal Astronomer, and he moved to the countryside close to the Royal Court at Windsor. There he was able to indulge his two passions: building telescopes, and investigating the nature and distribution of the stars. As most astronomers at the time were chiefly interested in the Moon and planets, Herschel had the field almost entirely to himself.

One of his projects was to investigate the large-scale distribution of stars in space. He was also fascinated by the nebulae – indistinct, fuzzy objects whose nature was then unknown. He received a copy of the first catalogue of nebulae prepared by the Frenchman Charles Messier, who had listed them as objects to be avoided when searching for comets. Herschel not only identified all the objects on Messier's list, but went on to discover over 2000 more nebulae. Among these were oval or ring-shaped nebulae which looked a little like the discs of planets: Herschel called these 'planetary nebulae', and the name still holds (although we now know that they are the remains of dying stars).

Herschel's holy grail was to discover the distance to a star. He reasoned that if he could find two stars that were almost directly in line, then the nearer of the two stars would appear to move relative to the other as the Earth travelled around the Sun. The extent of the shift would reveal how far away the star was.

Herschel spent many years studying these 'paired' stars. In some instances, he began to pick up a systematic motion that did not appear to be caused by the orbit of the Earth about the Sun. He realized instead that here he was observing a *genuine* pair of stars in space, and that the two bodies were in orbit about each other. These binary stars offered the first chance to test if gravity operated in the same way outside our Solar System as in it.

Although Herschel was the first to discover double stars (and we now know that more than half of all stars are double), his findings made him uncertain of his 'grindstone' model of the Galaxy. One of the basic tenets of the model was that all stars had the same brightness. However, it was obvious from looking at pairs of stars that their brightnesses covered a wide range. This discovery eventually led him to abandon the model altogether.

regions that provided the Aborigines with their celestial 'emu'. Herschel assumed that these rifts were actually holes in space, through which he could see through our local distribution of stars into empty space beyond.

Although Herschel could not measure the size of his grindstone, he could at least make an estimate of its dimensions. He expressed its size in terms of 'siriometers' – multiples of the distance to his standard star, Sirius. Herschel calculated that the grindstone measured 1000 siriometers across, and 100 siriometers thick. We now know Sirius to be slightly less than nine light years away, and so Herschel's Galaxy was about 9000 light years wide by 900 light years deep – about one-tenth the actual size of the Galaxy.

The grindstone model, drawn up before the end of the eighteenth century, bears a remarkable resemblance to the Galaxy as we know it today. But later in his life, Herschel became dissatisfied with it. The dark rifts nagged him: how was it that so many of them pointed directly at the Sun? Then he learned

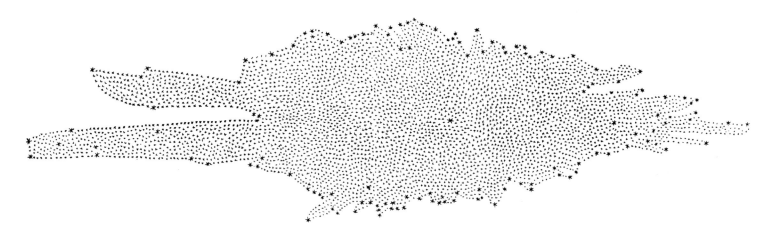

Herschel's method of 'stargauging' – plotting the 3-D distribution of stars from their apparent brightnesses – led to this, the first map of the Galaxy. The 'grindstone' model was, in fact, surprisingly accurate. But Herschel abandoned it later in life, because he thought that too many dark, empty 'tunnels' in space (rifts at edge of diagram) lined up on the Sun. We now know that the 'tunnels' are obscuring matter lying in the plane of the Galaxy.

that one of his basic assumptions was flawed. From his work on double stars, he discovered that stars are not equally bright – they have a wide range in intrinsic luminosity. Herschel also made a detailed study of the nebulae – faint, fuzzy patches in the sky. His work made him conclude that they lay *inside* our Galaxy, rather than being other star-systems outside.

And so, right at the end of his life, Herschel was to record: 'I must really confess, that by continuing my sweeps of the heavens, my opinion of the arrangement of stars and their magnitudes and some other particulars, has undergone a gradual change . . . For instance, an equal scattering of stars may be admitted in certain calculations, but when we examine the Milky Way, or the closely compressed clusters of stars, this supposed quality of scattering must be given up . . . We surmised the nebulae to be no other than clusters of stars disguised by their very great distance, but a longer experience and a better acquaintance with the nature of the nebulae will not allow a general admission of such a principle'.

Herschel actually renounced the grindstone model in his closing years. And his influence was so strong that other astronomers lost faith in it too. By the beginning of the nineteenth century, then, our ideas of the Galaxy were back in the dark ages once again. An eminent astronomer of the time, J. E.

Gore, wrote: 'Sir William Herschel's disc theory, as it is termed, was abandoned by its illustrious author in his later writings and is now considered to be wholly untenable by nearly all astronomers who have studied the subject'.

Herschel's ideas were replaced by a rather inelegant compromise, described here by his son John – himself a very distinguished scientist. 'Our situation as spectators is separated on all sides by a considerable interval from the dense body of stars comprising the galaxy, which, in this view of the subject would come to be considered as a flat ring of immense and irregular breadth and thickness . . .'. In other words, there were basically *two* systems of stars. One comprised the Sun and the stars we can easily see in the sky. The other was the distant ring of the Milky Way, separated from the local stars by a yawning gulf of empty space.

By the middle of the nineteenth century, however, William Herschel's original ideas were beginning to resurface. Their revival was thanks to the work of another great amateur astronomer – the third Earl of Rosse. Lord Rosse, who lived at Birr Castle, Parsonstown (now Birr), in Ireland, was a great devotee of large telescopes. He started off by building a 36-inch (0.9-metre) diameter telescope, and when that was completed, he decided to aim for something twice the size. It was to be

The 'Leviathan of Parsonstown' – Lord Rosse's mighty 72-inch telescope – was, for over half a century, the largest telescope in the world. Restricted to 'nodding' up and down the sky in its massive, castle-like mounting, the 72-inch nevertheless gave superb images of the 'spiral nebulae'.

the world's largest telescope, with a mirror 72 inches (1.8 metres) across.

Not everyone approved of the earl's ambitions. His fellow-countryman Sir Robert Ball, Professor of Astronomy at Dublin and a great popularizer of the subject, wrote: 'I think that those who know Lord Rosse well would agree that it was more the mechanical processes incidental to the making of the telescope which engaged his interest, than the actual observations with the telescope when it was completed. Indeed, one who was well acquainted with him believed that Lord Rosse's special interest in "the great telescope" ceased when the last nail had been driven into it'.

Lord Rosse proved his critics wrong. He began observing with 'The Leviathan of Parsonstown' in 1845, 'when the work was sufficiently advanced to make use of the instrument without personal danger'. He aimed the great telescope immediately towards those mysterious, faint, fuzzy patches in the sky – the nebulae.

The nebulae fascinated astronomers of the mid-nineteenth century. William and his son John Herschel had catalogued thousands of them in both the northern and southern hemi-spheres. But no-one knew what they were. Were they made of stars – enormous star-systems a long way away? Or were they gaseous – perhaps nearby stars or planets in formation?

The enormous light-grasp of Lord Rosse's telescope revealed the nebulae as never before. Many of them showed incredibly complex structure, which Rosse recorded in detailed drawings. He was particularly fascinated by those which were spiral in shape. As more and more spiral nebulae were found, astronomers began to wonder if the star-system that we live in was a lens-shaped, spiral nebula too. Had William Herschel's first theory been right after all?

In 1885, an answer of sorts arrived. In one of the largest, and presumably, closest of the spiral nebulae – the Andromeda Nebula – a star suddenly appeared. It rose to magnitude 6 on the astronomers' scale of brightness – just visible to the unaided eye. It was clearly an eruptive, or exploding, kind of star, and astronomers at that time knew of only one kind: a nova. By comparing the brightness of 'Nova Andromedae' with the intrinsic brightness of a normal nova, astronomers were able to calculate the distance to the nebula. The answer was that it lay comparatively close to us. The noted astronomer

Lord Rosse's enormous telescope revealed the structure of the spiral nebulae – like M51, here – for the first time. This beautiful drawing is incredibly accurate, and is worth comparing with a modern photograph of the galaxy. But at the time, these detailed observations only helped to fuel the controversy as to what the spiral nebulae were.

Andrew Ainslie Common said at the time: 'It is difficult to imagine that such an enormous object, as the Andromeda Nebula must be, is not very near to us; perhaps it may be found to be the nearest celestial object of all beyond the Solar System'. So the 'spiral nebulae' appeared to be local. The mystery of the shape and nature of the Milky Way still remained.

But astronomers were becoming uncomfortable. There were just too many loose ends around. By the turn of the century, several astronomers harboured gut feelings that our Milky Way *was* a spiral star-system, and that the spiral nebulae were independent systems in their own right. What was lacking was proof.

In a story with many twists and turns, the strangest manoeuvre was about to occur. The scene now changes to Peru, in 1908. Here – under the skies of the southern hemisphere – was an outstation of the Harvard College Observatory. From this southerly site, astronomers could photograph objects that didn't rise above the horizon in North America. Among them were two 'star clusters' (now known to be galaxies) called the Large and Small Magellanic Clouds. As it

turned out, the Small Cloud provided just what was wanted for a young researcher's project.

Henrietta Leavitt was interested in variable stars: stars that changed in brightness for one reason or other. What she needed was a sample of variable stars that were all at the same distance from us – even if she didn't know the actual distance – and the Small Cloud gave her just that. Using plates taken in Peru, Leavitt managed to identify 2400 variable stars in the Small Cloud. Most of them seemed to change in brightness in the same way: a steep rise to maximum brightness, followed by a more gradual fall-off. But the stars differed in the time they took to complete their periods of variation, and in their brightnesses.

The brighter a star was, Leavitt discovered, the longer it took to vary. Conversely, the fainter the star, the shorter was

Henrietta Leavitt's pioneering work identifying Cepheid variable stars in the Small Magellanic Cloud led to the first measurements of distances to galaxies outside our own – thus solving the problem of the nature of the 'nebulae'. In 1925, she was nominated for a Nobel Prize – but she had died four years previously at the early age of 51.

its period. The faintest stars completed their cycles in a day or so, while the brightest ones had periods of almost a month. When Leavitt compared the shape of the stars' characteristic 'light curve' to those of stars closer to home, she found that it exactly matched the light curve of the star delta Cephei – the prototype of the 'Cepheid' variable stars.

Leavitt went on to demonstrate that the Cepheids in the Small Magellanic Cloud could – in principle – be used as 'standard candles' to measure distances in space. The method itself was relatively straightforward: because Cepheids of the same intrinsic brightness had the same period of variability, then the scatter in their *apparent* brightnesses must be caused by their different distances. The problem was that the method was not calibrated. No-one knew how *intrinsically* bright any of the Cepheid stars were; even the closest one, delta Cephei, lay too far away for its distance to be measured directly.

Meanwhile, over in Missouri, a young would-be journalist was having problems in getting into university. Harlow Shapley arrived at the University of Missouri, only to discover that the school of journalism – which he was proposing to enter – had not yet been built. He returned the following year to find that nothing had changed. Shapley, with only $200 in his pocket, reasoned that it might be sensible to stay on in Missouri and find a different course to take. He started going through the courses catalogue alphabetically, and – as he wrote in his autobiography – 'there I was, all dressed up for a university education and nowhere to go. "I'll show them", must have been my feeling. I opened the catalogue of courses and got a further humiliation. The very first course offered was: A.R.C.H.A.E.O.L.O.G.Y . . . and I couldn't pronounce it (although I did know roughly what it was all about). I turned over a page and saw A.S.T.R.O.N.O.M.Y – I could pronounce that – and here I am!'

After Shapley had completed his undergraduate degree at Missouri, he went on to Princeton University. Here, he started to undertake research into eclipsing binary stars – systems of double stars whose light output changes as the two stars eclipse one another. His work was so good that it attracted attention at the Mount Wilson Observatory in California, then the leading astrophysical observatory in the world. In 1907 it had acquired a 60-inch (1.5-metre) telescope, which was the most modern at the time, as well as the world's largest (Lord Rosse's telescope was no longer in working order). George Ellery

Harlow Shapley, pictured here as a young man, used the then most powerful telescope in the world – the 60-inch reflector on Mount Wilson – to establish the scale of our Galaxy. By using Cepheid variable stars to measure distances to the globular clusters that surround the Milky Way, Shapley calculated that the 'skeleton' of our Galaxy was 300 000 light years across – so large that he believed it comprised the whole Universe.

Hale, who had set up the observatory, personally invited Shapley to come to Mount Wilson, to use the 60-inch for whatever he wanted to do.

Shapley's work on eclipsing binaries triggered an interest in genuinely variable stars, and he decided to use his time on the 60-inch to find out what made Cepheid stars tick. Although Henrietta Leavitt had done a thorough job on the way in which Cepheids varied, Shapley wanted to know *why* they changed in brightness. He discovered that the stars actually swell and shrink – a result of an instability just below the surface. Knowing how these stars pulsated enabled Shapley to work out the size of a Cepheid of a particular period as compared with the Sun. From here, it was but a short step to calculate its intrinsic brightness. By comparing a Cepheid's apparent brightness with its calculated intrinsic luminosity, you could find out how far away it was. Shapley was at last in a position to measure distances into the deepest realms of space.

For his distance-finding programme, Shapley cast around for prolific sources of Cepheids. He found one particularly rich seam amongst the globular clusters – dense balls of old stars that seemed to be a long way away. Shapley found that 60 out of 69 globulars had Cepheids in them. When he measured the distances to them, Shapley discovered two important things. First, the distribution of the globulars was spread equally above and below the plane of the Milky Way, which made Shapley conclude that they formed part of the same system. But some of the globulars were distributed much farther out than the stars of the Milky Way. Shapley suggested that the globulars formed a sort of outer skeleton to the Milky Way, with the most distant globulars marking its furthest bounds. Relative to our position, the distribution of the globulars was very lopsided. If they were really symmetrical about the centre of the Milky Way, then our Sun must lie a long way off-centre.

Shapley's relegation of the Sun to the edge of the Milky Way system did not provoke a lot of opposition amongst astronomers. But his measurement of the size of the system did. He calculated that the 'skeleton' of globular clusters was 300 000 light years across – ten times bigger than any previous estimate of the size of the Milky Way. It was so big that Shapley and his supporters proposed that the Milky Way and its surrounding skeleton made up the whole Universe.

Other astronomers strongly disagreed. One – Heber Curtis of the Lick Observatory – refused to believe that the Cepheids could be used as 'standard candles', and so disputed Shapley's whole scale of distances. Curtis instead believed that we live in the centre of a much smaller galaxy, and that the spiral nebulae were actually other galaxies.

Shapley fought back. The spiral nebulae, he argued, could not be other galaxies. If each was as large as our star-system – 300 000 light years across – the Universe would have to be absurdly large to contain them all. Further, he pointed out, his colleague Adriaan van Maanen had been photographing some of the spiral nebulae over the years and had definitely seen signs of rotation. If these measurements were correct (and we now know that they were not), and the spiral nebulae were as far away as Curtis believed, they would have to be spinning faster than the speed of light.

Eventually, in 1920, Shapley and Curtis were brought together at the National Academy of Sciences in Washington to fight it out. Their confrontation was called the 'Great Debate'. However, astronomers who were present remember it as being nothing of the kind; the whole thing appears to have been a pretty mild affair. Curtis backed his case with deep astrophysical ideas as to why the Universe was the way it was, while Shapley – unused to speaking in public – was so nervous that he couched his arguments at the simplest possible level.

In the event, there was no real debate, and at the end of it all no-one was really any the wiser. Those who were present left with a feeling that we live in a star-system with a definite shape, although whether it was the whole Universe, or just one star-system of many, was still in dispute.

One thing, at least, was certain: more observations were needed. And once again, Mount Wilson came up with the wherewithal. In 1917, the observatory had unveiled a new eye on the sky: the great 100-inch (2.5-metre) Hooker reflector, then the biggest telescope ever built. But the telescope suffered appalling teething troubles. The first mirror 'blank' – the unfigured lump of mirror glass – had huge bubbles in it. When George Ellery Hale, the observatory's founder, saw it, he suffered a nervous breakdown. A second glass disc, cast at the same factory in France, broke as it was cooling. Hale had a second nervous breakdown.

Eventually, the mirror was figured and fitted into the telescope. Hale's official biography graphically describes the night

The galaxy M87, at the heart of the Virgo Cluster, is a typical giant elliptical. Its million million stars are mostly old and red, and there is no dust and gas in the galaxy to create another generation of stars. M87 is surrounded by over 3000 globular clusters, which can be seen as fuzzy spots in this image. Using the latest technology, astronomers are currently trying to pick out Cepheids in these clusters, in order to establish a very precise distance to galaxies in the Virgo Group.

Until the early 1920s, astronomers still argued about the nature of the 'spiral nebulae'. Were they gaseous nebulae within our Galaxy, as Shapley's camp believed? Or true 'extra-galactic' star-cities, which Heber Curtis and his supporters maintained? Now we know that the 'spiral nebulae' lie well outside our Milky Way, and are galaxies in their own right. This photo shows a small group of galaxies in the constellation Leo.

For over 30 years, the 100-inch 'Hooker' telescope on Mount Wilson was the largest in the world. Its enormous light-grasp at last allowed astronomers to solve the problem of the spiral nebulae. This fresh perspective on the 'nebulae' later helped astronomers to unravel the structure of our own Galaxy.

when Hale and Walter Adams, an astronomer on the staff at Mount Wilson, went up to use the telescope for the first time. They 'climbed the long flight of narrow black iron steps to the observing platform. On the floor below, where a dim red light glowed, the night assistant pushed the control buttons . . . The observing platform rose and turned; the dome, its slit wide open to the star-filled sky, revolved in the opposite direction. The telescope itself turned until it pointed to the brilliant planet Jupiter.

'As soon as the telescope was set on Jupiter, Hale crouched down to look through the eyepiece, desperately eager to know if all the years of effort had been successful. He looked; and said nothing. Only the expression on his face told of the horror he felt. Adams followed. His expression was the mirror of Hale's. They were appalled by what they had seen. Instead of a single image, six or seven overlapping images filled the eyepiece'.

In a situation like that, there's only one thing to do: go away and see if things improve when the mirror cools down to the ambient temperature. So 'agreeing to meet three hours later, they went to bed. Hale lay down without undressing, but he could not sleep. An hour later he got up and tried to read a detective story, but this too failed. At 2.30 a.m., he returned to the 100-inch dome. Before long, Adams arrived, and confessed that he too had found sleep impossible. Once again, they climbed the long flight of steps to the dome floor, and then the narrow flight to the observing platform.

'By this time, Jupiter was out of reach in the west. They swung the great telescope over to the brilliant star Vega. Almost afraid to look, Hale again crouched down and looked into the eyepiece. He let out a yell. The yell told Adams all he wanted to know. The telescope was an unqualified success'.

Onto the scene now strides a young man who was also an unqualified success – as a lawyer. Edwin Hubble was a Rhodes

Edwin Hubble – photographed here with his ever-present pipe – laid the foundations of modern cosmology. His distance measurements to nearby galaxies, made with the 100-inch Hooker Telescope (background of shot), proved that they lay well outside our Milky Way. Later, Hubble would go on to classify the galaxies, and discover the expansion of the Universe.

Because Walter Baade was not an American national, he was excluded from serving in the Second World War – a situation he turned to his advantage by using the blackout to make ultra-deep surveys of space. His work with the 100-inch, and later the 200-inch, telescopes led to the discovery that there were two types of Cepheid, which had previously been confused. His researches put the distance scale on a much firmer footing, and the announcement of his findings – in 1952 – doubled the size of the Universe.

Scholar at Oxford, and very much a polymath. He had the supreme confidence that he could tackle anything he turned his hand to – and that included tank driving and amateur boxing. But he had also dabbled in astronomy courses when he was at college, and his interest grew so much that astronomy eventually replaced his dazzling legal career. 'I would much rather be a second-rate astronomer than a first-rate lawyer', he once said. 'All I want is astronomy'.

Mount Wilson quickly snapped Hubble up. His particular interest was the nebulae, and he started a programme of work with the 60-inch telescope to find out whether they were made of stars or gas. The results were clear: some, at least, contained stars.

To study these stars in detail, Hubble required the mighty light-grasp of the 100-inch. In 1923, he found that some of the stars in a rather nondescript nebula known as NGC 6822

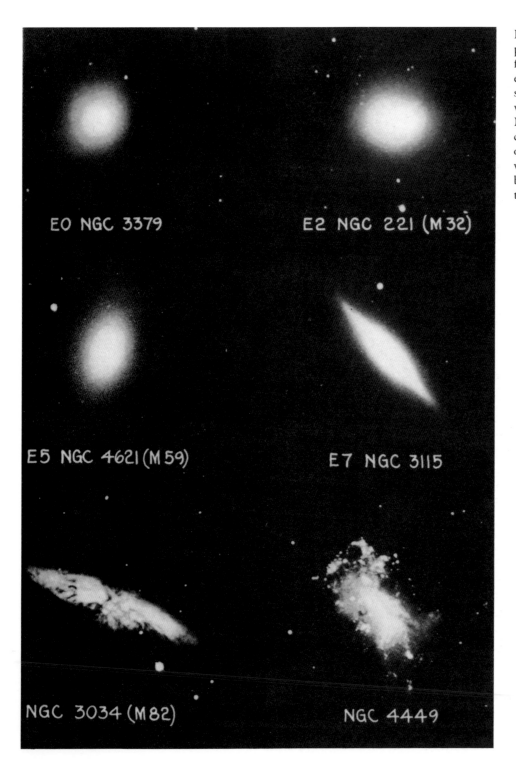

EO NGC 3379

E2 NGC 221 (M32)

E5 NGC 4621 (M59)

E7 NGC 3115

NGC 3034 (M82)

NGC 4449

Edwin Hubble chose these hand-labelled photographs to illustrate his pioneering classification of galaxies. Ellipticals come with different degrees of flattening, from circular (E0) to spindle-shaped (E7). To Hubble, any galaxy without symmetry was 'irregular', so he lumped NGC 4449 with M82 (which we would now call a starburst galaxy). Hubble's classification of spirals has withstood the test of time well, with its division into ordinary spirals (S) and barred spirals (SB), each subdivided according to the openness of its arms, from Sa/SBa to Sc/SBc.

Sa NGC 4594   SBa NGC 2859

Sb NGC 2841   SBb NGC 5850

Sc NGC 5457(M101)   SBc NGC 7479

BOX 2. **How the Universe doubled in size – overnight**

The stars used as 'standard candles' to plumb the depths of space – Cepheid variables – are amongst the most luminous stars known. Named after the star delta Cephei, a typical Cepheid has a temperature that varies between 6000 and 7500 degrees Celsius, and an average luminosity 10 000 times that of the Sun. Cepheids are at an advanced stage of their lives – at a point where they are beginning to run out of fuel – and this leads to them becoming unstable. During this phase, the star expands and contracts, and it may change in size by as much as 30 per cent. The bigger (and brighter) the star, the longer each cycle takes: hence the period–luminosity relation.

Edwin Hubble's measurements of the distances to Cepheid variables in some of the 'nebulae', in the 1920s, placed them well outside our Milky Way. It was this discovery that made astronomers realize that our Milky Way was just one galaxy amongst many others. However, not everyone was totally confident that the Cepheids were giving the right answers. One baffling anomaly was that most of the galaxies appeared to be very much smaller than our own. Some were just one-tenth its size; and even the great Andromeda Galaxy was apparently only one-sixth as large.

Was our Galaxy really a giant among galaxies? The answer came in a roundabout way. In 1945, the astronomer Walter Baade was lucky enough to use the 100-inch Mount Wilson telescope under ideal conditions: when the lights of Los Angeles were blacked out because of the Second World War. He was able to make a thorough examination of the stars in the Andromeda Galaxy, and reached the conclusion that they were divided into two 'populations'. Red and yellow (Population II) stars lived in the centre of the galaxy and in the 'outer skeleton' halo; hot blue and white (Population I) stars resided in the spiral arms.

Seven years later, Baade got even luckier: he was one of the first observers on the great '200-inch' telescope on nearby Palomar Mountain. He was confident that the huge telescope would show him RR Lyrae variable stars – rather fainter stars than Hubble's Cepheids – in the halo of the Andromeda Galaxy. He was baffled that none showed up. Instead, he found a new kind of Cepheid variable that appeared to live in the halo only. These stars were 1.5 magnitudes fainter (four times fainter) than the Cepheids in the main body of the galaxy – the ones that Hubble had used to derive its distance.

From his work on star populations, Baade realized that there must be two types of Cepheid corresponding to the different populations. But Hubble and his colleagues had been unaware of this. By comparing Population I Cepheids in the Andromeda Galaxy with the fainter Population II Cepheids in the Milky Way, they had come up with a distance (and correspondingly, a size) that was much too small. We now know that the Andromeda Galaxy lies two and a quarter million light years away, and is half as large again as the Milky Way.

In fact, Baade's 1952 discovery that there were 'classical' Cepheids and fainter Population II Cepheids literally doubled the size of the Universe at a stroke. In the succeeding years, astronomers refined their measurements of galaxies more distant than Andromeda, and found that they were more remote than even Baade had thought. This meant that spiral galaxies were all considerably larger than Hubble had calculated.

At the same time, other research was showing that Harlow Shapley had overestimated the size of our Galaxy: instead of being 300 000 light years across, its diameter was only about 100 000 light years. These two discoveries, taken together, had important repercussions on our understanding of the Milky Way Galaxy. Instead of the Milky Way being ten times larger than most spirals, it turned out to be average in size – a reassurance to astronomers who like to think that we do not live in an especially privileged position in space.

were changing in brightness. These stars were Cepheid variables – the key to measuring a galaxy's distance – but Hubble did not realize this important fact immediately. Instead, he went back to Andromeda. Here, other astronomers had recently found several very faint novae, some 20 000 times dimmer than the nova of 1885. If it were these fainter stars that were similar to novae in the Milky Way, then Andromeda must be far beyond our own star-system.

BOX 3. **Classifying the galaxies**

Edwin Hubble's term 'extragalactic nebulae' was unwieldy, but it was a phrase that he chose carefully. As part of his distance-measuring project, he also systematically photographed the nebulae in order to categorize them. Although many were spiral – as Lord Rosse had discovered – not all of them were. And although the spiral nebulae clearly contained stars, those that had spherical or elliptical shapes seemed to have none (we now know that they do). Hubble believed that they were wholly gaseous – rather like the Orion Nebula on a much larger scale. This is why he preferred to call them all 'extragalactic nebulae' – literally, 'nebulae outside our Galaxy'.

On the other side of America, Harlow Shapley was also working on these objects. He and his colleagues at Harvard strongly objected to the term 'extragalactic nebulae', and much preferred the shorter 'galaxies'. It was not until the death of Hubble in 1953 that east and west coasts of the USA were finally united in agreeing on 'galaxies'.

Hubble decided to classify the galaxies on the basis of their appearance on photographic plates. This was because he believed that the different shapes of the galaxies represented different phases in a galaxy's evolution. For this, Hubble drew closely on the work of the English astronomer Sir James Jeans, who had worked theoretically on how a gas cloud would collapse as it formed into stars. This, believed Hubble, was the key to understanding the nature of the galaxies.

Hubble's spiral galaxies came in three types, which he classified according to the tightness of their spiral arms. Type Sa had large centres and tightly wound arms; Sb galaxies were intermediate; and Sc types had small central bulges and uneven, open spiral arms. He noted a parallel sequence – the barred spirals – that had bar-shaped centres (SB), and so could be divided into SBa, SBb and SBc.

The galaxies without spiral arms came in two kinds. Some had no real shape at all – like the Large and Small Magellanic Clouds – and Hubble defined them as 'irregular'. Others were spherical or elliptical in shape, and seemed to have no stars in them. These Hubble classed as 'elliptical' (E), followed by a number to quantify the extent of their ellipticity (from 0 for circular in outline, up to 7 for very elongated). Intermediate between ellipticals and spirals were S0 (or lenticular) galaxies, which were lens-shaped objects rather like the bulge of an Sa galaxy without any arms.

Hubble published his original theory on the nature of the galaxies in 1922, and expanded on it in 1925. He proposed that a galaxy began life as a spherical gas cloud, then flattened out and developed spiral arms as it rotated. In time, the spiral arms slowly unwound. Finally, a galaxy would end its days as an Sc or even an irregular galaxy.

In later years, Hubble himself began to doubt that this was an evolutionary sequence. Astronomers now know that it is not. Each type of galaxy has quite distinct characteristics. For instance, elliptical galaxies are made up almost entirely of old red stars – a result of rapid and super-efficient star-birth. Irregulars have few old stars, but lots of gas – a sign of *inef-ficient* star-formation. Spirals contain a mixture of old and young stars, along with plenty of gas to make up future generations. Unless externally disturbed, galaxies start off life as one of the Hubble types, and stay that way. However, it is an immense tribute to Edwin Hubble that his original classification scheme has proved so invaluable, and has weathered the passage of time so well.

In the autumn of 1923, Hubble began to photograph the Andromeda nebula night after night, searching for novae. He struck lucky immediately. In Hubble's own words: 'The first good plate in the program . . . led to the discovery of two ordinary novae and a faint object which was at first presumed to be another nova'. Hubble looked for this star on older photographic plates stored at Mount Wilson, and these 'established the faint object as a variable star and readily indicated the nature of the variation. It was a typical Cepheid with a period of about a month, and hence its absolute luminosity at maximum . . . was about 7000 times as bright as the Sun. To appear as faint as the observations indicated, the required distance was of the order of 900 000 light years'.

Spurred on by this success, Hubble took a closer look at the

BOX 4. **Naming the heavenly bodies**

This book – like all other guides to astronomy – is littered with a bewildering assortment of names for stars, nebulae, pulsars and all the other denizens of our Galaxy. These have come about through the ages as astronomers use what – at the time – appears to be a logical system of names.

The constellations have names derived, by and large, from Greek legend, such as the story of how the king and queen of Ethiopia, Cassiopeia and Cepheus, had a daughter called Andromeda who was rescued from the ravaging sea-monster Cetus by the hero Perseus. All these characters are represented by constellation patterns. The ancient Greeks could not see the stars around the south pole of the sky, and later astronomers filled in this region with constellations such as the toucan (Tucana), Table Mountain (Mensa) and the pendulum clock (Horologium).

From ancient times, the brightest stars have had individual names, such as Rigel, meaning 'the foot [of Orion]' in Arabic, and Arcturus, which is Greek for the 'bear-warden' because this star follows the Great Bear. Sometimes there have been mistakes in transliteration: Betelgeuse was originally 'yad al-jawza', the hand of Orion, but the present form is closer to the Arab word for armpit!

In 1603, the German astronomer Johann Bayer gave the brightest stars Greek letters, constellation by constellation, with the star pattern designated in the genitive form. Betelgeuse thus became alpha Orionis. England's first Astronomer Royal, John Flamsteed, was even more systematic, giving numbers – constellation by constellation – to all the stars visible to the naked eye. For fainter stars, we usually resort to their numbers in the more recent catalogues, such as the Henry Draper (HD) which contains 225 300 stars.

Variable stars have their own idiosyncratic system. They are named by constellation, in order of discovery, in a sequence that begins with R, S, T . . . After Z, the names begin again with RR, working through to RZ, then SS to SZ; after ZZ, the sequence carries on with AA, through to QZ. This accounts for 334 stars: if a constellation contains more variables, they are simply V335, V336, and so on.

'Fuzzy patches' in the sky cover a variety of objects: including star clusters, nebulae and galaxies beyond our own. Because they all look similar through a telescope, this diverse collection of objects shares the same kind of catalogue designations. The most obvious have their own names, like the Pleiades, while others bear names from star catalogues, such as omega Centauri and 47 Tucanae. Astronomers have more recently christened many more with fanciful nicknames, such as the Ring Nebula, the North America Nebula and the Trifid Nebula.

The first catalogue of fuzzy patches came from the Frenchman Charles Messier, in the late eighteenth century. Messier was the 'ferret of comets', and his catalogue was intended as a check-list of objects that he might confuse with comets. Messier's catalogue contained 103 entries, now shown by a prefix 'M', such as M42 (the Orion Nebula). A century later, the Danish astronomer J. L. E. Dreyer compiled a list of 7840 fuzzy objects, in his New General Catalogue, updated a few years later by two Index Catalogues. These 'NGC' and 'IC' numbers are still the most widely used.

After the Second World War, researchers began to find sources of radio waves in the Cosmos. At first, they could not pin down the locations precisely, and the objects were named by constellation: Cassiopeia A, for example was the strongest radio source in that ancient star-pattern. Later surveys were more precise, and new sources became known by their catalogue numbers – often in the pioneering third Cambridge ('3C') survey.

X-ray astronomers followed a similar route two decades later. The strongest X-ray source in the constellation of the Swan is thus Cygnus X-1. Astronomers now think that it contains a black hole. The fainter Cygnus X-3, lying in the same direction but much further away, is a powerhouse of cosmic rays.

This system brought a considerable amount of confusion, because one star or nebula could end up with six or more names. The remains of the supernova that was seen to explode in 1054 is variously M1, NGC 1952, Taurus A, 3C144 or Taurus X-1 – but most people know it by the nickname 'the Crab Nebula'.

Now, astronomers working at all wavelengths use a much more rational system. They simply name a source by its coordinates in the sky – like locating a house by giving its Ordnance Survey reference. The first number refers to the right ascension (celestial longitude) and the second to the declination (celestial latitude). A prefix may reveal – rather confusingly – either the telescope used to discover the object or the kind of object it is. Thus PKS refers to the Parkes radio telescope, whereas PSR means pulsar.

Astronomers working near the Galaxy's centre use a variant on this system. They refer their coordinates to the Galaxy's equator, so the first number is now the galactic longitude and the second the galactic latitude, with the letter 'G' in front. Even these apparently logical systems can have their problems – as witness the title of a paper in *Nature* in August 1991: 'Unusual interaction of the high-velocity pulsar PSR 1757−24 with the supernova remnant G5.4−1.2' – yes, they are both parts of the same exploded star!

variable stars in NGC 6822. They, too, were Cepheids, and indicated that this nebula lay some 700 000 light years away. Another nebula, M33, turned out to be at a similar distance.

Hubble's measurements showed that these nebulae must lie outside our local system, which Shapley had estimated to be 300 000 light years across. Hubble had rushed his first observations – and the undeniable conclusion that Andromeda lay outside the Milky Way – to Shapley. On opening the missive, Shapley had commented to his colleague Cecilia Payne-Gaposhkin 'here is the letter that has destroyed my Universe'. Now there was no doubt: these were all star-systems independent of our own. Hubble called them 'extragalactic nebulae' (which we now refer to as 'galaxies' with a lower-case 'g'), and he went on to study them in detail.

In a sense, Hubble discovered our Galaxy: he was able to show that it was an independent star-system in its own right. But – back in the 1920s – he could say next to nothing about its size, its structure, or its make-up. Other galaxies – some of them millions of light years away – were no problem; their layout was on view for all to see. The unravelling of the geography of *our* Galaxy was to take many more years, and a great deal of ingenuity. Later chapters in this book tell of the fascinating – and often challenging – quest to map out and explore the highways and byways of our local city of stars.

Charles Messier (1730–1817) – painted here at the age of 40 by Desportes – was nicknamed 'the ferret of comets' by King Louis XV. Although he discovered 16 comets, he is best remembered now for drawing up a catalogue of objects that the unwary astronomer might take to be a comet. Most of the bright star clusters and nebulae in our skies – and many galaxies – are known by their 'M' numbers.

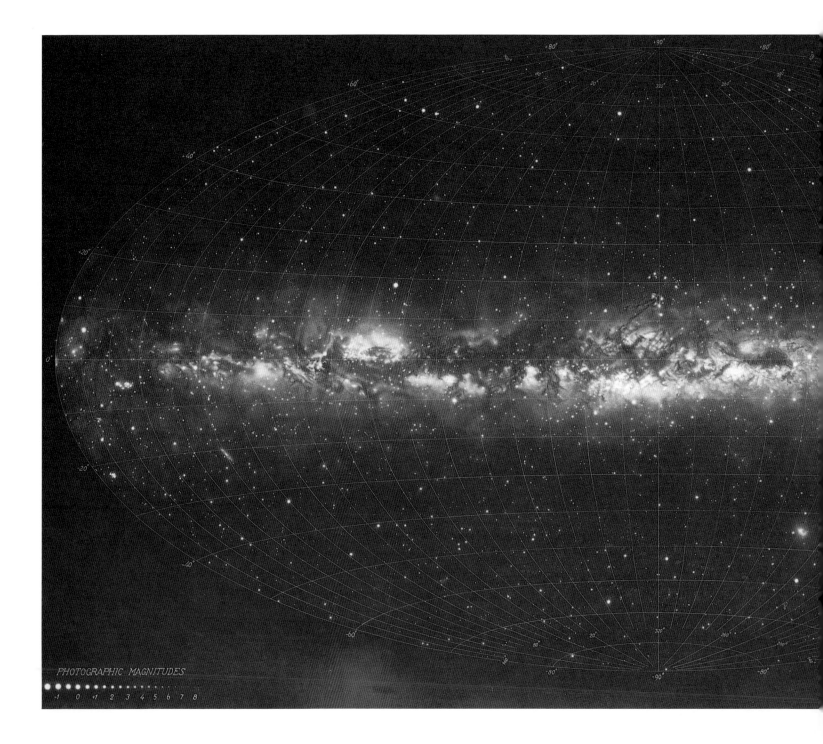

LUND OBSERVATORY

MARTIN KESKÜLA
TATJANA KESKÜLA

This photographic mosaic of the Milky Way, compiled at the Lund Observatory in Sweden, reveals the clumpiness of the stars making up our Galaxy. The dark regions are huge clouds of obscuring material in which stars will later form. The two 'detached portions' at lower right are our two closest galaxies – the Large and Small Magellanic Clouds.

CHAPTER 2

The Local Group

One of life's more wonderful experiences is to stand outside on a warm, scented summer's night in the southern hemisphere. Above, the stars crowd the sky from horizon to horizon, making the whole dome of heaven seem luminous and ablaze. And amongst the stars – high up in the sky – you'll spot two large, misty patches of light, looking for all the world like fine-weather clouds lit by the Moon. They certainly made an impact on Antonio Pigafetta, the official recorder of Ferdinand Magellan's first circumnavigation of our globe. In 1521, he suggested that they be called the Clouds of Magellan in honour of the great navigator, who had died during the long journey. To this day, we know them as the Large and Small Magellanic Clouds.

The chill nights of a northern autumn, although scarcely as alluring, provide an equally fascinating vista. Hovering above the thin line of stars supposed to represent the chained maiden, Andromeda, is a faint, oval blur. Like its southern counterparts, this indistinct smudge is a remote collection of millions of stars – a galaxy in its own right. The Andromeda Galaxy is the furthest object visible to the unaided eye: it lies two and a quarter million light years away.

The Magellanic Clouds are nearer than Andromeda, at about 170 000 light years – but they are still well outside our own Milky Way. These three star-systems are among our Galaxy's nearest neighbours in space. Together with a couple of dozen much smaller galaxies, they make up our 'Local Group': our home cluster of galaxies.

Clusters of galaxies, rather than isolated galaxies, are the rule in the Universe. Some clusters contain thousands of members, but our small Local Group is probably much more typical. Unlike the giant clusters which are roughly spherical, the Local Group is flattened in shape, and measures about five million light years across. The next nearest galaxies lie several million light years beyond.

Seen from a respectable distance – say, through a small telescope from a hypothetical planet in the Virgo Cluster – our Local Group would excite little interest. An observer would see only the Milky Way, the Andromeda Galaxy and the small spiral galaxy M33, which lies close to Andromeda.

Of these three, the Andromeda Galaxy is much the largest, and the natural place to begin a tour of the Local Group. This galaxy is almost half as big again as the Milky Way, and it contains twice as many stars – an estimated total of 400 billion.

In his *Book on the Constellations of the Fixed Stars*, published in about AD 964, the Arabic astronomer Abu I-Husain al-Sufi describes a 'nebulous spot' lying at the mouth of a big fish that the Arabs superimposed on the figure of the chained maiden Andromeda. Although this is the first definite reference to the Andromeda Galaxy, the Roman poet Festus Avienus wrote a tantalizing line about the chained constellation in the fourth century AD: 'thin clouds tie her arms with twisted knots'. Did this reference to 'clouds' mean that the Romans were familiar with the misty outline of the Andromeda Galaxy?

In 1612, the German astronomer Simon Marius became the first person to view the Andromeda Galaxy through a telescope – and it left him little the wiser. It looked, he wrote 'like the flame of a candle seen through horn'. A century-and-a-half later, Charles Messier included it in his catalogue of nebulous objects as number 31.

A copy of Messier's catalogue came into the hands of William Herschel, in England. With his powerful telescopes, Herschel thought – mistakenly – that he could almost make out individual stars in the nebula. The true shape of Andromeda was only revealed as recently as 1888, when the English astronomer Isaac Roberts took the first photographs. They revealed what is not at all obvious to the eye observing through a telescope – that the 'nebula' is spiral in shape.

The problem in seeing the true glory of this giant spiral is that it is presented to us at a very steep angle – 78 degrees. The long-exposure photographs pioneered by Roberts show clearly the spiral arms in the galaxy's outer regions. But even then it was not clear what this object – and the other 'spiral nebulae' – actually were. Many astronomers thought that they were nearby gas clouds condensing into stars, each with a planetary system.

As we saw in the previous chapter, the problem was solved in the 1920s, by Edwin Hubble, at the Mount Wilson Observatory in California. The investigations by Hubble and his successors into Andromeda were also very important in helping astronomers to get a feel for the contents and the 'geography' of our own Milky Way. The main difference, apart from the smaller size of the Milky Way, is that Andromeda has rather more tightly wound arms: in Hubble's classification it is an Sb galaxy, whereas astronomers believe that the Milky Way is intermediate between an Sb and an Sc.

Hubble's spiritual successor was Walter Baade, a German-

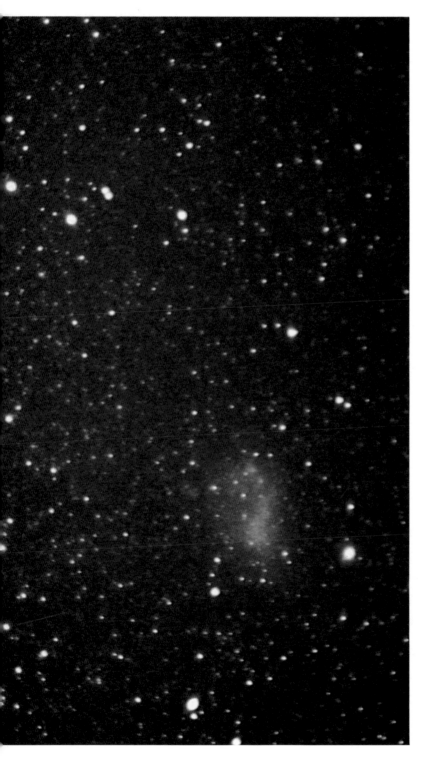

The Large Magellanic Cloud (upper left) and the Small Magellanic Cloud (right) are among the most striking sky-sights in the southern hemisphere. Our closest galactic neighbours – they are probably satellites of the Milky Way – both galaxies are also members of the Local Group. This excellent amateur photograph, which captures the galaxies' appearance in the sky so well, was taken with a 50 millimetre lens.

Contrast-enhanced image of the largest galaxy in the Local Group, M31 in Andromeda. This unusual view highlights the outer spiral structure of the galaxy, and shows up its companions, M32 (left) and NGC 205 (right). Our Milky Way would look rather like this from outside, but M31 is half as large again.

This radio image of the Andromeda Galaxy reveals the distribution and movement of cool hydrogen gas in its spiral arms. Mapped at a wavelength of 21 centimetres with the Westerbork Synthesis Radio Telescope in the Netherlands, the brightness of the gas shows where it is most densely concentrated, while the colours – ranging from blue for approach to red for recession – show its motion. The image vividly reveals the rotation of the galaxy, with the lower arms approaching, and the upper arms receding.

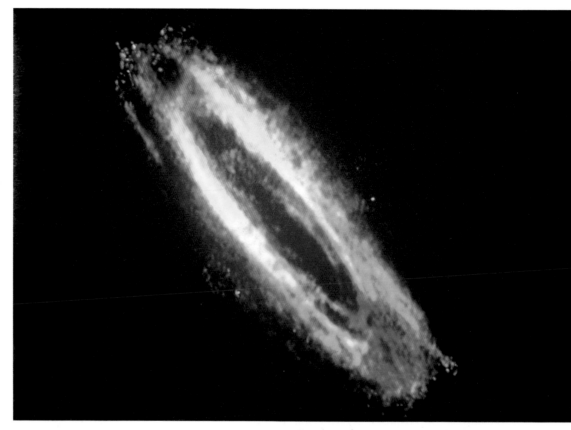

born astronomer who was invited to move to the United States in 1931 to use the large telescopes in California. During the war-time blackout, Baade was able to study individual stars in the Andromeda Galaxy in detail. He found that they fall into two classes: the spiral arms contain hot young and middle-aged stars (Population I), while the central regions comprise much older stars (Population II).

From 1948, Baade was able to use the newly completed '200-inch telescope' – formally known as the Hale Telescope – on Palomar Mountain, also in California. His detailed dissection of Andromeda has proved a cornerstone to understanding all spiral galaxies (including our own).

Spiral galaxies have four distinct parts. The component that shows up spectacularly in photographs is the pair of arms: there are usually two arms, though some galaxies have four and others display a whole collection of short arm segments.

The spiral arms are studded with the galaxy's gems – sapphire-blue hot, young stars and ruby-red nebulae. These are the youngest objects in the galaxy, and astronomers often call then 'extreme population I' objects.

The arms are the most prominent feature of the second of the galaxy's components, the disc. This is a thin, flattened region of stars that surrounds the galaxy's central regions, like the white of a fried egg. The disc is made up of young and middle-aged stars, corresponding to Baade's Population I, and contains almost all the gas and dust that reside in the galaxy.

In the centre of the disc is a bulge of yellowish-orange stars, resembling the yolk of the fried-egg (even down to its colour!). Surrounding the lot is the fourth component, a vast spheroidal halo, of dim red stars. Both the bulge and the halo consist of old, Population II stars.

The halo is an ancient fossil remaining from the galaxy's

early days, and so contains unique records of a galaxy's past. Unfortunately, it is the least obvious part of any galaxy. Whereas the disc has plenty of bright stars in its mix, the halo consists only of dim stars. Some of the halo stars, however, give themselves away by their gregariousness. These are stars that reside in globular clusters, great balls each containing as many as a million stars. Astronomers have managed to log 600 globular clusters in the Andromeda Galaxy. (This is another indication of the superiority of Andromeda to our Galaxy, which can muster only 140 globulars.)

Whereas astronomers have to search for the older and dimmer parts of a galaxy, the younger regions – the disc and spiral arms – show up in any photograph. In the case of Andromeda, these stars form a disc almost 150 000 light years across. But the galaxy is actually larger still. About one-tenth of its mass is in the form of gas, and astronomers can tune into the unmistakeable signature of cold hydrogen gas threading between the stars by using radio telescopes tuned to a wavelength of 21 centimetres.

These observations show that the gas exists not only between the stars: the disc of hydrogen actually extends far beyond the bounds of the visible galaxy – out to a total diameter approaching 200 000 light years. This gas isn't spread uniformly through the galaxy: most is concentrated into a ring some 40 000 light years from the centre, where the majority of the star-forming regions also lie.

Nor does the gas behave in an orderly fashion. Instead of orbiting around the centre of the galaxy in a straightforward, circular manner, the gas indulges in some fairly spectacular dynamics. The inner arm on the north-east part of the galaxy – as well as racing around in its orbit – is also falling in towards the galaxy's centre at a speed of 100 kilometres per second (about 200 000 miles per hour!) At present, the reason isn't known, but astronomers suspect that the pull of one of Andromeda's companion galaxies could be responsible.

Although the disc of Andromeda contains many young stars, there are surprisingly few large regions of star-birth – as we find, for example, in the less massive galaxies M33 and the Large Magellanic Cloud. Instead, there are many smaller nebulae – nearly 1000 in all – marking the maternity wards of new stars.

We know, from studying other spiral galaxies, that nebulae tend to follow the line of the spiral arms. Several astronomers have used the nebulae to try to trace the spiral arms in Andromeda, although the steep angle at which we view Andromeda drastically foreshortens the perspective and the true picture is not clear cut. Most astronomers accept that the nebulae outline two spiral arms that wrap around the galaxy more than once, in such a direction that they trail as the galaxy rotates. But there's evidence for much confusion and distortion in the arms – again a result, perhaps, of warping by the tidal pull of M32, one of Andromeda's companion galaxies.

The Andromeda Galaxy's central bulge is smooth and featureless – but not without its excitements. First, it contains (or perhaps we should say, contained) the most famous star to exist in the galaxy: S Andromedae. This star literally burst onto the astronomical scene in August 1885, and was first reported by E. Hartwig at the Dorpat Observatory in Estonia. By the end of the month, it was just visible to the naked eye. Astronomers of the time assumed it was an ordinary nova (an eruption that we now know to involve the nuclear explosion of gases heaped onto one star by a companion), which typically brighten to some 100 000 times the brilliance of the Sun. As we saw in the previous chapter, this misinterpretation led astronomers to conclude that S Andromedae, and hence the Andromeda Galaxy (or 'Nebula' as it was then called), lay quite close to us.

But when Edwin Hubble, in the 1920s, established that Andromeda was actually very remote, the truth about S Andromedae began to dawn. At that distance, it had to be much brighter than a nova. In fact, S Andromedae turned out to be a different kind of beast altogether. It was a supernova – a star that catastrophically detonates at the end of its life with the brightness of almost a billion Suns. Later research has shown that spiral galaxies like Andromeda typically produce one supernova every 50 years, so we are well overdue for S Andromedae's successor!

The other area of interest is right in the heart of the central bulge: the galaxy's core. In 1993, the sharp eye of the Hubble Space Telescope showed that the centre is double: it consists of two star clusters five light years apart. The fainter cluster is the galaxy's core. In 1987, two groups of astronomers studied in detail the speed at which stars orbit the core, and came to the conclusion that they are being held in orbit by an enormous – but unseen – mass. The prime candidate is a black hole, weighing in at 10 million to 100 million times the mass of

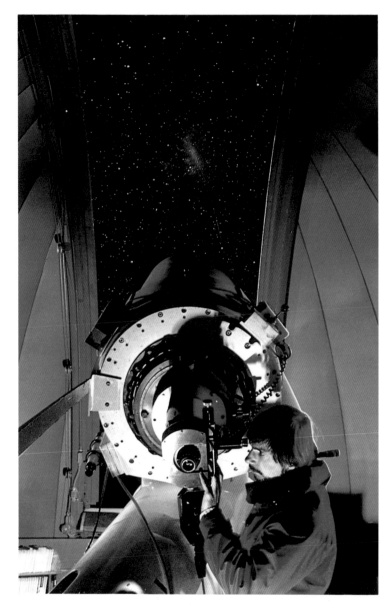

Canadian astronomer Ian Shelton adjusts the 61-centimetre telescope at Las Campañas Observatory in Chile, with which he discovered Supernova 1987A in the Large Magellanic Cloud. The LMC (with the supernova at the lower left edge of the galaxy) can be seen through the open dome. Shelton discovered the supernova on 24 February 1987, but – unusually – it took until mid-May to reach its maximum brightness.

the Sun. Such a lurking monster may be a feature of all giant galaxies, including our own – but it lurks unseen until it is fed! The brighter cluster seen by the Hubble telescope may be the remains of a small galaxy being chewed up by the Andromeda giant – to have survived complete disruption, it may also contain a massive black hole.

Even a small telescope shows that Andromeda is not alone in space: it has two small companions, M32 and NGC 205. The Milky Way also has two companions, the Magellanic Clouds, but there is an important difference. The Magellanic Clouds are irregular galaxies, containing lots of gas and dust. Andromeda's two companions are elliptical, made mainly of old stars.

A French astronomer, Guillaume Le Gentil, first noticed M32 in 1749, and his fellow countryman Charles Messier included it in his famous list of nebulous objects. Oddly enough, Messier's list did not include Andromeda's second small companion, even though we know that he saw it, because it is included in a sketch that Messier made of M31 and M32. Although some historians have numbered this galaxy retrospectively as 'M110', most astronomers use the later catalogue number NGC 205.

As seen from our vantage point, M32 practically adjoins the Andromeda Galaxy. It is nearly circular in shape, and only 6000 light years across. With just two billion stars, it is only half of one per cent as massive as the Andromeda Galaxy itself. Yet even this small galaxy has been able to wreak havoc on its host. The spiral structure of the Andromeda Galaxy appears to have been considerably distorted by M32. But it also appears that Andromeda has managed to get its own back, for M32 shows signs of having once been a much larger galaxy that has been stripped of most of its stars.

Andromeda's other companion, NGC 205, is twice the size of M32. Although it is not a lot brighter than M32, it makes up for any deficiency by being very unusual. To begin with, it is distorted by the tidal pull of Andromeda, which causes a 'twist' in its outermost regions. And – as far as elliptical galaxies go – it displays one remarkable oddity. As well as its normal population of old red stars (Population II), NGC 205 contains some hot blue stars, cool gas and a prominent band of dust – all the trademarks of the regions of young Population I stars that we find in spiral galaxies. Perhaps the galaxy is rejuvenated by its proximity to Andromeda, which may keep it topped up

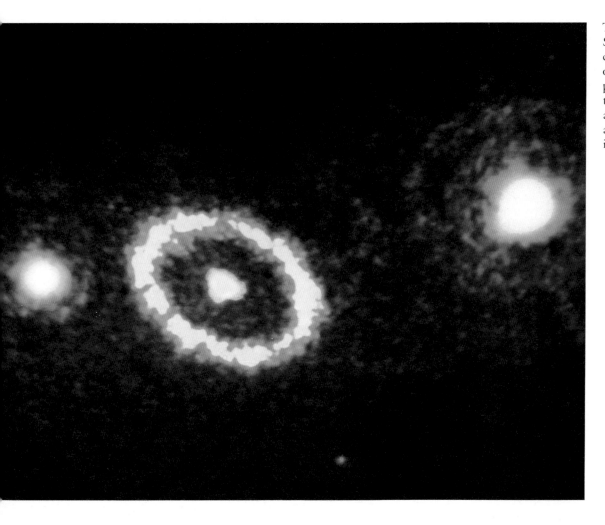

Three and a half years on from the explosion of Supernova 1987A, the Hubble Space Telescope captured an image that demonstrates the pace of change. The original star has vanished. In its place (centre) is a cooling mass of tightly knotted debris, colour-coded red. Surrounding it is a ring of gas (coded yellow), which will eventually disintegrate when the expanding debris hits it. (The two stars to the left and right are not associated with the supernova.)

with gas. This explanation, however, cannot account for the star-forming regions in a rather similar, but further-flung, satellite galaxy of Andromeda, NGC 185.

This galaxy and NGC 147 form a pair of rather distant and faint companions to Andromeda. The giant spiral has eight small companion galaxies in all, the four faintest discovered as recently as 1970. All are ellipticals, bar one – which may in fact be simply a detached part of the Andromeda Galaxy itself.

Not far from the Andromeda Galaxy in the sky is the third largest member of the Local Group, M33. Although it, too, is a spiral, M33 and Andromeda form a very dissimilar pair of twins. Andromeda is large and bright, but is reticent about

displaying any spiral structure: its arms are tightly wound and largely hidden by its tilt. In contrast, M33 may be small and dim, but it makes a beautiful display of fine spiral arms that we see face-on.

Charles Messier found this galaxy in August 1764. It lies in Triangulum, a constellation that borders Andromeda, and some astronomers call it the 'Triangulum Galaxy'. But, in contrast to its more famous neighbour, the nickname has never caught on, and plain 'M33' it remains in most astronomy books.

If you have a really dark sky, and good eyesight, you may just be able to make out M33 with the naked eye. It is certainly

BOX 1. **Supernova of a lifetime**

In the early morning of 24 February 1987, Canadian astronomer Ian Shelton was ready for bed after spending three hours photographing our nearest neighbour galaxy, the Large Magellanic Cloud. To his annoyance, however, the photograph looked as though it had a serious fault: a blob that looked rather like a star – *very* like a star, in fact. He checked with other astronomers at the Las Campañas Observatory in Chile. Night assistant Oscar Duhalde said that he had indeed spotted a new star in the Large Magellanic Cloud earlier that evening. Meanwhile, across the Pacific Ocean, amateur astronomer Albert Jones was amazed as this star unexpectedly blazed through the eyepiece of his telescope.

Shelton, Duhalde and Jones had discovered a supernova: the brightest exploding star to be seen for 383 years – since the invention of the telescope, in fact. And in 1987, astronomers had not only powerful optical telescopes to study the supernova, but instruments to pick up all kinds of radiation from gamma rays to radio waves, and even the elusive particles known as neutrinos.

Because Supernova 1987A lay in another galaxy, almost 170 000 light years away, it was not as brilliant as some of the nearby supernovae in our Galaxy, such as those recorded in 1604, 1572 and 1054. But its location had some advantages. First, it was not obscured by the dust that hides distant stars in the plane of our Galaxy. It lay in a well-known galaxy, so astronomers knew its distance instantly. And the Large Magellanic Cloud has a slightly different composition from the Milky Way, so the supernova gave astronomers new insights into the way its composition affects the life and death of a star.

Astronomers had also – for the first time – photographed the star before it exploded. Catalogued as Sanduleak −69° 202, it had been a blue giant star, twenty times as heavy as the Sun and fifty times larger. The star's size came as a surprise. According to theory, a star about to go supernova should have swollen to become a red supergiant several hundred times the size of the Sun. With hindsight, astronomers believe that Sanduleak −69° 202 did swell to become a red supergiant. At this stage, it puffed off a ring of gas that grew to become 1½ light years across. Compared with its siblings in the Milky Way, however, this star had less of the heavy elements that absorb energy welling up from its centre, so it shrank again to blue-giant size.

Deep within the core of Sanduleak −69° 202, trouble was brewing. Nuclear reactions had converted the material here to nuclei of iron, the most stable of the elements. Increasing pressure and temperature hammered these nuclei together harder and harder, until they all suddenly began to disintegrate. The star's core collapsed on itself. The temperature soared to 50 000 million degrees, creating a vast flood of neutrinos, which shot through the star and into space at the speed of light. The immense outward pressure from the neutrinos combined with shockwaves around the core to rip the outer layers of the star off into space. Supernova 1987A was born.

The neutrinos carried away 99.99 per cent of the supernova's energy – sufficient to power all the stars in our Galaxy for several years! On the morning of 23 February 1987, the flood of penetrating neutrinos shot through the Earth. A tiny fraction collided with atoms on the way through our planet. In Japan and the United States, particle detectors in two large tanks of water picked up the signals of just 19 neutrinos. These confirmed, for the first time, the theory that the collapse of a star's core causes a supernova explosion.

A few hours later, the light from the explosion reached the Earth. Because Sanduleak −69° 202 was smaller than most exploding stars, the supernova did not become as bright as astronomers expected at first. But a surprise was in store. Instead of fading again, the supernova became gradually brighter. By the end of May, it was 250 million times more brilliant than the Sun.

This energy must have come from radioactive elements created in the explosion. American astronomer Stan Woosley calculated at the time that the shock wave would have changed 0.07 solar masses of the star's material into radioactive nickel-56, which decays quickly into cobalt-56. The radioactive cobalt decays over a period of months into stable iron-56. In the process, it can supply exactly the energy needed to keep the supernova shining. Astronomers confirmed this theory when they picked up spectral lines from cobalt, at both visible and gamma-ray wavelengths.

By the early 1990s, the exploding star was no more than a dim cloud of gas, one per cent as bright as the star was before the explosion. But astronomers then began to pick up an increasing amount of radio emission. This radiation was generated as the expanding gases from the explosion began to crash into the previously ejected gas ring. The sites of old supernovae in our Galaxy are marked by strong sources of radio waves and X-rays, so called 'supernova remnants'. By the year 2000, Supernova 1987A should have a second blaze of glory – at radio and X-ray wavelengths – as it turns into 'Supernova Remnant 1987A'.

an easy sight in low-powered binoculars. But if you turn a high-powered telescope on to it, the galaxy disappears! Its light is spread out so thinly that there is very little contrast between the galaxy and the background sky. A high magnification thins out the light even more, and the contrast becomes so slight that the human eye cannot detect the galaxy at all.

When Edwin Hubble first used Cepheid variables to measure the distances to galaxies in the 1920s, M33 was one of his first targets. He found it was at about the same distance as the Andromeda Galaxy. Although astronomers now put the Andromeda Galaxy further away than Hubble did, his conclusion that they are at the same distance from us should hold. A more recent study of these Cepheids at infrared wavelengths also indicates that the two galaxies are the same distance from the Milky Way.

But the astronomer who took over the mantle of Baade at Palomar Observatory, Allan Sandage, disagrees. He has looked again at Hubble's original data on the Cepheids, and measured them again with modern techniques. This indicates that M33 lies well beyond Andromeda, at something like three million light years from us. So, it may be somewhat surprising to learn, astronomers are still arguing over the distance of one of the brightest galaxies in the sky.

M33 is considerably smaller than either the Andromeda Galaxy or the Milky Way – only 40 000 light years in diameter. It contains only one-tenth as many stars as Andromeda. But it makes up for this paucity by a vivid display of huge bright nebulae. The largest has its own number in the New

General Catalogue – NGC 604 – and is much larger and more brilliant than any nebula in the Andromeda Galaxy, and possibly any in the Milky Way.

These nebulae are powered by newly born stars, and M33 has a real production line for them: despite its smaller mass, it is producing new stars at a similar rate to the Andromeda Galaxy. The young stars and nebulae are strung along the open arms that characterize an Sc spiral in Hubble's classification. Near the centre, there seem to be two main arms; but further out these break down into half a dozen or more separate curving strings of stars and nebulae.

And the centre of M33 holds a mystery. In 1978, American astronomers launched the Einstein Observatory into orbit around the Earth, to look for X-rays coming from space. The observatory found that the centre of M33 is a prolific source of X-rays, which can switch on and off rapidly. Despite this outpouring of X-rays, the galaxy's core does not shine brilliantly at any other wavelength: in particular, radio telescopes do not pick up any radiation. Astronomers know of many distant galaxies that have 'active nuclei' – probably a disc of hot gas around a massive black hole – but these generally emit radiation at a whole range of wavelengths. If M33 harbours a black hole in its centre, gobbling up spare gas in its vicinity, it has a unique method of spewing out the resulting energy in the form of X-rays alone.

Although the Andromeda Galaxy and M33 are two of the three most eye-catching galaxies in the Local Group (along with the Milky Way), they are far from being 'typical'. These three giants are outnumbered ten times over by dwarfs – galaxies so small and dim that we can only see them at all because they are right on our extragalactic doorstep.

This is a sobering reminder of what the Universe is really made of. When we look out into the depths of space, we see many spiral and large elliptical galaxies. But if our Local Group

FACING PAGE Computer-enhanced image of M33, the third-largest galaxy in the Local Group. This photograph is actually a combination of separate images taken through red, green and blue filters, that were later combined by computer to reveal the actual – but enhanced – colour of the galaxy. The nuclear bulge is dominated by the light of old red and yellow stars, while the outer arms shine the characteristic sapphire blue of young stars and gas. Dust along the inner edges of the arms – a great reservoir of material for future star-birth – appears rusty-red.

BOX 2. **Eyes on the invisible**

Astronomy is a strange kind of science. Researchers cannot take their subject matter into the laboratory and experiment on it. They have to rely on what nature puts on offer and then observe it from a distance. Traditionally, astronomers have depended on the light that comes from stars, nebulae and galaxies. But even the most powerful optical telescopes, backed up with electronic light detectors, can only tell us a limited amount about the Universe.

Light is only a small part of the whole electromagnetic spectrum. It reaches from gamma rays – with wavelengths less than one-billionth the length of waves of light – to radio waves that are a billion times longer. When we observe the Universe at optical wavelengths only, we are in the position of someone listening to a symphony with ears that are sensitive only to middle C and the two immediately adjacent notes.

Some objects in our Galaxy in fact emit very little light, but produce copious amounts of radiation at other wavelengths. The gaseous remains of supernovae, for example, only show up well at radio and X-ray wavelengths. We also have to tune to non-optical wavelengths if we want to investigate the more distant parts of our Galaxy, because light is absorbed by ubiquitous particles of dust in the space between the stars: infrared and radio telescopes have, for example, revealed a maelstrom of activity at the centre of the Galaxy. These telescopes have also laid bare the centres of the dense dark clouds where stars are being formed.

Gamma rays come from the most energetic objects we know: gas falling into black holes, the compact and rapidly spinning pulsars and hyper-energetic cosmic rays crashing into gas in space. The most brilliant X-ray sources consist of gas being torn from a star and falling towards a black hole or a pulsar. Other X-ray sources, such as supernova remnants, consist of gas at a temperature of a million degrees Celsius and more. We pick up ultraviolet radiation from hot stars and from the superheated atmospheres of ordinary stars.

Ordinary light comes mainly from stars with surfaces at a temperature of several thousand degrees: it is no coincidence that our eyes have evolved to pick up radiation from an average star, the Sun! Cooler stars, such as red giants, emit infrared radiation around ten times the wavelength of visible light. Longer-wave infrared comes from objects that are still colder, in particular, clouds of dust in space.

These clouds also contain molecules, which broadcast at characteristic wavelengths in the radio band. The most famous radio signal from space, however, comes from hydrogen atoms spread throughout the Galaxy: they produce radiation at a wavelength of 21 centimetres. Many radio astronomers do not tune to particular wavelengths, but pick up radiation that covers a broad spectrum. These waves come either from hot clouds of gas, or from fast electrons moving through a region containing a magnetic field and generating 'synchrotron radiation'.

Most of these radiations from space cannot pierce the Earth's mantle of air. Our atmosphere has only two wavelength 'windows': these allow light and radio waves to penetrate to sea level. This is why we can see the stars, and why radio astronomy was the first of the 'new astronomies' to develop, in the 1930s. Infrared waves can plunge far enough through the atmosphere for astronomers to pick them up with telescopes on the highest mountains, such as the 4200-metre peak of Mauna Kea in Hawaii. But the clearest views of the infrared sky require a telescope in space.

For the rest – ultraviolet, X-rays and gamma rays – we must put our telescopes above the atmosphere, preferably in Earth-orbiting satellites. These new astronomies only opened with the start of the Space Age in 1957, and satellites are constantly opening new frontiers in our understanding of the Galaxy and beyond.

is a typical part of the Cosmos, then by far the most common galaxies are the dwarfs.

Even when the dwarfs lie close at hand, it's not easy to spot them. A few are near enough and bright enough to have got into the New General Catalogue, but most have turned up either as a result of special searches or simply because a keen-sighted astronomer has spotted a faint smudge on a very long-exposure photograph of the sky. So, even now, it's hard to be

These galaxies may be unprepossessing to look at, but because they are so numerous – they are by far the commonest kind of galaxy in the Local Group – they may be the most typical in the Universe. Sex-B and Leo-I are both dwarf elliptical galaxies. Each contains roughly a million stars, but the stars are spread so thinly that these galaxies are very dim. Dwarf ellipticals are made almost entirely of old red stars, and have no material left for star-formation. But although they would appear boring, these galaxies may be the most important to study, for they hold the clues to the way that larger galaxies form.

certain that all the dwarf members of our Local Group have been found.

The dwarfs fall into two types. Dwarf ellipticals consist of old stars, with little or no sign of young stars, gas or dust. Dwarf irregulars contain gas, dust and glowing nebulae, crowned with a scattering of bright young stars.

Dwarf ellipticals are the smallest – and probably commonest – galaxies in the Universe. For all the world, they look just like flaws on a photographic plate. It's not just their small size – about 5000 light years in diameter – that helps them to evade detection. They lack brilliant young stars, and the dim stars they do contain are so spread out that the galaxies contrast very little with the background sky.

Although dwarf elliptical galaxies have a similar number of stars (about a million) to their close cousins, the globular clusters, they are much more extended. An idea of their sparseness can be gauged from the fact that the density of stars is only one-thousandth of that in the region around the Sun in our Galaxy. The stars are so spread out that alien astronomers in a dwarf elliptical would see only two or three stars in their night sky!

It's interesting that the other class of dwarfs are irregulars – galaxies without any particular shape. There are no dwarf spirals. The irregulars, however, contain the same mixture of stars and gas as a spiral. This indicates that a gas-rich galaxy will be an irregular if it is low in mass; it can 'grow' spiral arms only if it contains a certain minimum mass, at least 10 billion stars. The dwarf irregulars in the Local Group are well below this limit, generally consisting of 100 million stars or less.

One of the largest of the dwarf irregulars is NGC 6822, in the constellation Sagittarius. This was the galaxy in which Edwin Hubble first recorded Cepheid variable stars, paving the way to his later plumbing the depths of space. This galaxy contains something like two billion stars.

At the other end of the scale is a tiny irregular called GR8, which refuses to follow the rule that the smallest galaxies are ellipticals. GR8 is the smallest galaxy yet found. Shaped like a cosmic footprint, it measures only 1000 light years from 'heel' to 'toe'.

The two kinds of dwarf seem to have different distributions within the Local Group. Like tugs around a pair of liners, the

dwarf ellipticals swarm around the Andromeda Galaxy and the Milky Way. The irregulars stray further afield, preferring to mark the edges of the Local Group. It's not clear yet whether this difference is real: it may be that dwarf ellipticals are also common near the boundaries of the Local Group, but we cannot detect such faint galaxies at this distance.

The gaggle around the Milky Way comprises eight dwarf ellipticals, at distances ranging from 300 000 to 500 000 light years. These one-off discoveries have no complex catalogue names. Astronomers have named them after the constellation pattern in which they appear to lie: Fornax, Draco, Ursa Minor.

Harlow Shapley discovered the first of these galaxies, Sculptor, in 1938, on photographic plates taken with a telescope in South Africa. It appeared as a smudge, looking just like a fingerprint on the photograph. Shapley's team also found the Fornax galaxy, while four more turned up in the 1950s when a large camera – a Schmidt telescope – photographed the whole sky visible from Palomar Mountain. A southern version of this instrument, operating in Australia, turned up the Carina dwarf in 1977 and Sextans in 1990.

At first, astronomers were not sure whether these companions of the Milky Way were genuine galaxies in their own right, or simply globular clusters of our Galaxy's halo – albeit globular clusters whose stars were rather sparsely distributed. After all, these galaxies and the globular clusters contain similar numbers of stars of the same type: old Population II red stars. In addition, some of the Galaxy's bona fide globular clusters do stray as far as 300 000 light years from the centre of the Milky Way, out to the realm where the dwarf ellipticals are found.

But astronomers are now convinced that dwarf ellipticals are indeed a different breed from the globular clusters. Whereas the globulars are a by-product of the birth of our Galaxy, the dwarf ellipticals were born independently. One piece of evidence to back this view is the fact that the dwarf ellipticals seem to contain some stars that are only about two billion years old – far younger than the age of the Galaxy, which dates back at least 10 billion years.

Another important clue, according to American galaxy expert Paul Hodge, comes from the fact that the Fornax galaxy has six globular clusters surrounding it. None of the globular clusters in our Galaxy contains smaller globular clusters; indeed, there are good theoretical reasons why this could not have happened as the globulars formed around the young Milky Way. 'Dogs have fleas, but fleas don't have fleas', as Hodge puts it. 'Thus Fornax must be a true galaxy, as its six fleas testify'.

If the dwarf ellipticals were producing new stars two billion years ago, they must then have had copious reservoirs of gas and resembled their more showy cousins, the dwarf irregulars. Where has that gas gone? A rash of supernovae may have blasted the remaining gas out of the galaxies. This theory does not, however, explain why the dwarf ellipticals apparently cluster around the Milky Way (and the other large spiral, the Andromeda Galaxy). Instead, the original gas-rich dwarfs may have strayed into the outer parts of the Milky Way, where they ran into gas that makes up a vast tenuous atmosphere to the Galaxy. This collision stripped the gas from the smaller galaxies, to produce the skeletal dwarf ellipticals.

Scars of collisions and close galactic encounters are certainly evident when we study our Galaxy's two larger companions, the Magellanic Clouds. The stormy past uncovered by recent research forms a striking contrast to their serene appearance in the skies today.

These two misty patches were of course known to the inhabitants of the southern hemisphere long before any Europeans ventured onto the southern seas. The Australian Aborigines, for example, believed that the larger cloud was a portion of the Milky Way that had been ripped away, leaving the dark gap now known as the Coal Sack. Before Magellan's name was invoked, western navigators knew them as the 'Cape Clouds'.

In the 1750s, Abbé Nicholas Lacaille cemented this historic association. Most of the Large Magellanic Cloud lies in the constellation Dorado (the goldfish), but Lacaille formed some faint stars on its fringe into the constellation of Table Mountain (Mensa). The Large Magellanic Cloud sits on the edge of this constellation in the same way as the famous 'tablecloth' clouds drape Table Mountain itself.

FACING PAGE One of the first galaxies to have its distance measured, NGC 6822 is also the brightest of the dwarf galaxies in the Local Group. Although classified as an irregular galaxy, NGC 6822 looks as if it has some organization, rather like the Large Magellanic Cloud, with a bar of stars and a curving line of nebulae. This galaxy has only one-hundredth the mass of the Milky Way.

BOX 3. **Clusters of galaxies**

The Local Group is a typical corner of the Universe – a small cluster of galaxies bound together by the mutual gravity of its members. Nearby, we can see numerous other examples – for instance, the group around the galaxies M81 and M82 in Ursa Major, some ten and a half million light years away; and the Sculptor Group of galaxies. Something like 90 per cent of the galaxies in the Universe are thought to belong in small groups like these. These clusters are called 'irregular' – they have no particular shape, no real structure, and contain a mix of all kinds of galaxies.

'Regular' clusters are very different. These are giant swarms of thousands of galaxies, like the Virgo and Coma clusters. Their inhabitants even look alike – the hotch-potch of all galaxy types is replaced by the uniform appearance of hundreds of elliptical or S0 galaxies. Galaxy expert Paul Hodge describes the difference between irregular and regular clusters as being like the difference between sprawling Los Angeles and compact New York – even down to the regular blocks of samey sky-scrapers in Manhattan.

Why do regular clusters contain so few spirals? The answer may be that they have literally had the stuffing knocked out of them. Most regular clusters contain a great deal of hot gas, which emits copious X-rays. Almost certainly, this gas has been driven out of what were originally gas-rich galaxies by collisions over the lifetime of the cluster.

Collisions have apparently led to another bizarre feature seen in the regular clusters. Many of them are dominated by one or two supergiant galaxies that live right in the centre of the cluster. These 'cD' galaxies have grown to the size they are today – sometimes hundreds of times larger than an average galaxy – by devouring their companions. Situated at the centre of gravity of the cluster, these galaxies only have to sit and wait for unwary galaxies to swim into their domain – upon which they get eaten alive. Some cD galaxies contain several dense cores within their distended structure, which appear to be the shrivelled husks of galaxies they have consumed.

Although regular clusters make up very obvious clumpings of matter in the Universe, detailed galaxy counts reveal that even larger structures exist. On the biggest scales of all, clusters of galaxies bind together to make superclusters – enormous groupings of galaxies up to 200 million light years across. However, most of a supercluster is empty space. The clusters of galaxies tend to line up in thin filaments surrounding enormous voids, making space look like a Swiss cheese on its largest scales.

A typical supercluster is made of two or three regular clusters, with several irregular clusters mixed in. Our Local Group, along with many of the nearby irregular clusters, is part of the Virgo Supercluster, which is centred on the giant regular cluster in Virgo. The gravity of all the galaxies in the supercluster is sufficient – locally – to have a small effect on the expansion of the Universe and to influence the motion of our Galaxy in space.

Even a small telescope dissolves the mists of the Large Magellanic Cloud, breaking it up into a plethora of stars, star clusters and nebulae. When John Herschel – son of the great Sir William – published in 1847 a list of pioneering observations made from the Cape, he listed 919 interesting objects in the Large Magellanic Cloud alone. A leading Australian amateur astronomer, Ernst Hartung, has enthused more recently: 'the telescope discloses vast regions of star clouds, open and globular clusters, gaseous nebulae both compact and diffuse, and innumerable scattered stars exposed for observation in a setting which is unique in the whole sky'.

Photographs have shown this galaxy to be an even more

FACING PAGE The huge elliptical galaxies M84 and M86 dominate this image of the central regions of the Virgo Cluster of galaxies. Unlike our puny Local Group, the Virgo Cluster – which is about 60 million light years away – contains thousands of members. Many of them are giant ellipticals – a type of galaxy that the Local Group lacks. The Local Group and several other nearby groups of galaxies appear to be dominated by the gravity of the Virgo Cluster, and the whole conglomeration is known as the 'Virgo Supercluster'. Superclusters are the Universe's largest building blocks.

This long-exposure photograph reveals the faint outer extensions of the Large Magellanic Cloud, which, at 169 000 light years, is the nearest galaxy to our own. Although rather irregular looking, astronomers now classify the LMC as a spiral galaxy – the prototype of the small, often one-armed, Sm 'Magellanic spirals'. This image shows the galaxy's central bar, the giant star-forming region 30 Doradus (pink blob at top) and, below it, Supernova 1987A close to its maximum brightness.

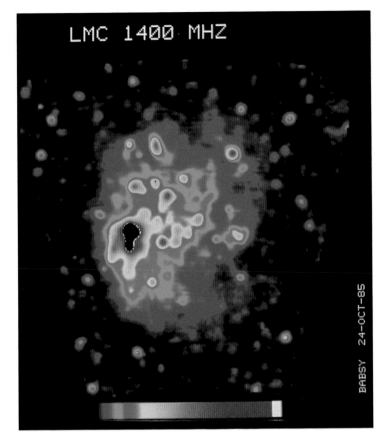

This colour-coded radio map of the LMC shows the locations of hot, gaseous nebulae where stars are currently forming (like the SMC, the LMC too had a burst of star-formation 100 million years ago). Observations at different radio wavelengths can reveal whether hot gas-clouds are nebulae where stars are being born, or supernova remnants where stars have died.

wonderful cornucopia, containing over 6000 star clusters and a similar number of nebulae. It consists of 20 billion stars – about one-tenth the content of the Milky Way – and a couple of billion solar masses of gas that are ready to turn into stars.

When Edwin Hubble developed his neat way of classifying galaxies (or 'nebulae', as he called them) he described both Magellanic Clouds as 'typical irregular nebulae – highly resolved, with no nuclei and no conspicuous evidence of rotational symmetry'. Despite this assertion, astronomers have now found some distinct family resemblances between the Large Magellanic Cloud and spiral galaxies of the barred type.

For a start, the stars near the centre form an elongated spindle shape that resembles a spiral galaxy's central bar. Long-exposure photographs show curved lines of stars and nebulae reaching out from the ends of the bar, resembling rudimentary arms. Clinching the case are measurements made in the 1980s. By measuring precisely the relative distances of the stars in different parts of the galaxy, astronomers have found that the Large Magellanic Cloud is a thin disc, rotating steadily around its central point.

The Large Magellanic Cloud is now regarded as the prototype of a new kind of spiral galaxy. The Sm category ('m' for Magellanic) is intermediate between an Sc and an irregular. It lies at a crucial frontier in mass: a heavier galaxy has enough gravity to corral its matter into well-defined arms, whereas the material in a less massive galaxy tends to go its own way, producing no overall pattern.

Long-exposure photographs certainly show that the Large Magellanic Cloud is much bigger than any genuine irregular galaxy. With a diameter of 35 000 light years, it is almost as broad as M33, the Triangulum Galaxy. Even as seen with the naked eye, the Large Magellanic Cloud covers about five degrees of sky – about ten times the diameter of the Full Moon. Long-exposure photographs reveal its fainter regions stretching to twice this size.

As the nearest galaxy to the Milky Way, and with its contents flaunted so ostentatiously to the view of telescopes in the southern hemisphere, the Large Magellanic Cloud has been a Rosetta Stone for astronomers studying the life history of stars. The most spectacular example was the flamboyant death of a star as a supernova seen in 1987. Among its other disclosures, the supernova has provided the most accurate distance yet to this closest of all galaxies, just 169 000 light years. But a more subtle investigation of its stars has provided a biographical insight into the past life of the Large Magellanic Cloud itself.

By measuring the brightness and colours of the stars within a cluster, astronomers can work out the cluster's age and so gain clues to a galaxy's activity earlier in its history. Our Galaxy formed massive globular clusters in its youth, and has created smaller open clusters at a roughly constant rate ever since. But the Large Magellanic Cloud spawned only a few clusters in its early days, and then halted its production of stars for eight billion years. Just a few billion years ago, the gas in the Large Magellanic Cloud began to condense into star clusters once more.

The new spate of clusters in the Large Magellanic Cloud consists not only of small open clusters, but includes many massive globular clusters as well. We can currently observe the construction of a million-solar-mass globular cluster in the centre of a giant nebula so brilliant that you can see it with the naked eye. Officially, it is called NGC 2070 or 30 Doradus, but its menacing tentacles of glowing gas have given rise to the nickname the Tarantula Nebula.

What made this galaxy suddenly decide to produce stars again, after such a long pause? The most likely answer is a close encounter with our Galaxy. Most astronomers believe that the Magellanic Clouds are true satellites of the Milky Way, lazily orbiting our Galaxy together in a period that takes billions of years to complete. When the Large Magellanic Cloud passed close to us some four billion years ago, so the theory goes, the gravity of our Galaxy stirred up the gases of our smaller companion and so prompted a new session of star-birth.

A recent near miss seems to have wreaked more severe damage on our second satellite galaxy, the Small Magellanic Cloud. As its name suggests, this galaxy has only one-quarter the mass of its larger companion and appears only half the size of the Large Magellanic Cloud in our skies.

Few astronomers would quibble that the Small Magellanic Cloud is an irregular galaxy. From our vantage point, it resembles a tadpole stretching across the constellation of Tucana (the toucan). But its true shape is remarkable. Over the past decade, astronomers have carefully measured the distances to the stars in the Small Magellanic Cloud. The nearest stars are 160 000 light years from the Milky Way and the farthest are 220 000 light years away. In three dimensions, then, the Small Magellanic Cloud is a long cylinder – 60 000 light years long and 15 000 light years in diameter – with its long axis pointing towards us. (An alternative interpretation, recently proposed by Australian researchers, is that the galaxy resembles a boomerang seen end on.)

Sometimes called the Tarantula Nebula (for obvious reasons!), the LMC's 30 Doradus star-formation complex is 900 light years across, and contains 500 000 solar masses of hydrogen gas. If it lay as close to us as the Orion Nebula, it would be visible during daylight. It was once thought that the stars forming inside it were freakishly massive supergiants, but higher-resolution observations have revealed that apparently massive 'single' stars are dense clusters of several, more normal stars.

FACING PAGE Slightly more distant than its larger neighbour the LMC, the Small Magellanic Cloud lies about two Galaxy-diameters away from us. The prominent spine of glowing nebulae running along the SMC is the result of a burst of star-formation that began about 100 million years ago – possibly triggered by a close approach to our Galaxy. This close encounter had an even more devastating effect on the structure of the SMC. It apparently ripped the galaxy apart, so that we are now looking end-on at a region of stretched-out debris.

A galaxy this shape cannot be stable. Most astronomers now believe that the Small Magellanic Cloud is being torn apart, and its component parts spread along our line of sight. The mighty force required to rip this galaxy to pieces must have been the gravity of our massive Milky Way, catching the Small Magellanic Cloud in an awkward way as it passed close by us between 100 and 200 million years ago.

Radio astronomers have tracked down further evidence of the destruction of the Small Magellanic Cloud. Stretching half-way round the sky is a narrow ribbon of hydrogen gas: at one end it merges into a large pool of hydrogen that surrounds both Magellanic Clouds. Astronomers have suggested a wide range of fanciful ideas to account for the long Magellanic Stream. Most likely, however, it is gas spilt from the disrupting Small Magellanic Cloud and spreading out along the orbit of our companion galaxies around the Milky Way.

Eventually, the gas in the Magellanic Stream – and perhaps other fragments of the Small Magellanic Cloud – will fall in towards the Milky Way. The remains of its satellite will help to swell, ever so slightly, the bulk of our own Galaxy.

CHAPTER 3

Geography of the Galaxy

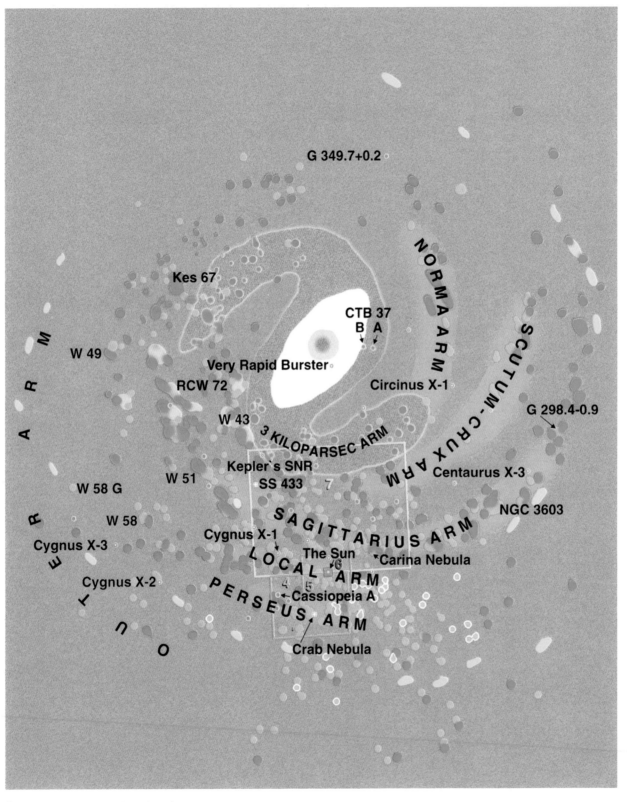

A

5,000 Light Years

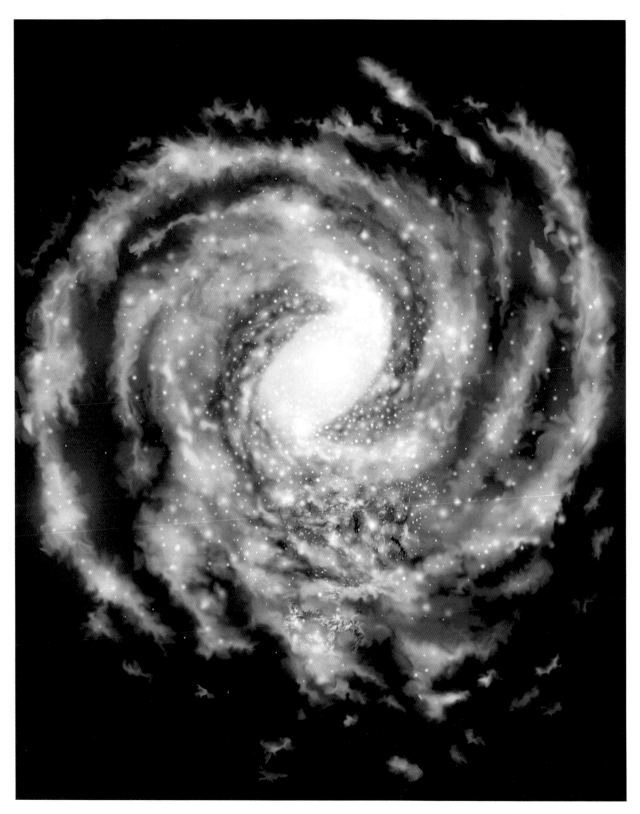

B   **KEY TO MAP A OVERLEAF**

KEY TO MAP A

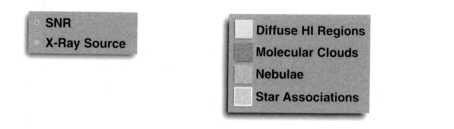

- ○ SNR
- ◉ X-Ray Source

- ▢ Diffuse HI Regions
- ▢ Molecular Clouds
- ▢ Nebulae
- ▢ Star Associations

MAP A An outline of the true shape of our Galaxy appears in a plot of the location of objects that trace its spiral arms and central bulge. The best tracers of the spiral arms are molecular clouds and nebulae – picked out at great distances by their radio emission. In the outer regions, large clouds of hydrogen atoms (HI) can be located unambiguously, while nearer to the Sun optical astronomers can pick out nebulae and young star clusters and associations. The barred centre is delineated by the distribution of red giant stars, derived from infrared observations.

The yellow boxes indicate the regions covered in more detail in later chapters. Some of the most prominent individual objects in the Galaxy are also marked (the Sun is shown for reference only).

MAP B A bird's-eye view of the Milky Way shows our Galaxy as realistically as possible in the light of current observations. This structure has been built up by fleshing out the known tracers of the spiral arms and nuclear region, with no attempt to force the spiral arms to fit any mathematical pattern.

To an astronomer living in the Andromeda Galaxy, our Milky Way Galaxy would hold few mysteries. A backyard stargazer could tell immediately how bright the Milky Way is, how large it is and its Hubble type – the pattern that its stars present to the Universe. With a professional-standard telescope, an astronomer in Andromeda could instantly lay bare the spiral pattern of the Milky Way and check whether its centre is undergoing a spasm of activity. Other instruments would show the main regions of star-birth, and the most powerful sources of cosmic rays and other high-energy radiation.

For astronomers on Earth, it has taken decades of unstinting work just to begin to unravel these basic facts about our Galaxy – even after the early pioneers had established that we live in a star system that is akin to Andromeda. Immersed right among the Galaxy's stars, we are in the position of not being able to see the wood for the trees. Astronomers suffer the added complication that most of the wood is obscured in fog – thick palls of interstellar dust.

But in recent years, researchers have at last been able to answer most of these basic questions about our home Galaxy. The breakthrough has come from careful observations at all wavelengths from gamma rays to radio waves. These have allowed us to 'see through' the obscuring dust, and to measure the distances to stars, nebulae and gas clouds throughout the Galaxy. By delving a bit deeper into the stars' spectra, astronomers have even been able to work out a broad-brush picture of the history of the Milky Way.

There has been one unspoken guiding principle. Once astronomers in the 1950s had satisfied themselves that the Milky Way was a fairly medium-sized 'star-city', it became obvious that our Galaxy must be – in general terms at least – similar to the many thousands of other galaxies in the Universe. Galaxies of this mass come in only two forms. The Milky Way was certainly not a fat gas-free elliptical galaxy, so it must be a spiral – a sibling to the other two spiral galaxies in the Local Group, the giant Andromeda Galaxy and the lightweight Triangulum Galaxy, M33.

The Milky Way has turned out to be in many ways a cross between its two neighbours. Whereas Andromeda shines with the light of 40 billion Suns and the Triangulum Galaxy is barely one-tenth as bright, the Milky Way has the brilliance of 15 billion Suns. And the Galaxy's diameter of 100 000 light years is almost an exact average between the sizes of Androm-

eda and M33. In terms of temperament too, our Galaxy lies between the slumbering Andromeda, where there is very little activity of any kind, and the fiery Triangulum Galaxy, with its glowing nebulae and X-ray-powered core. Busy regions of star-birth in the arms of the Milky Way and disturbances in its centre make our Galaxy an interesting – but not too dangerous – place to live.

Like other spirals, we would expect the Milky Way to consist of a thin disc of gas, dust and bright young stars, marshalled into some kind of spiral shape, a swarm of older stars forming a central bulge, and a faint surrounding halo, composed of the very oldest stars.

Ironically, the dim halo of our Galaxy was the first part to be recognized in anything like its modern form, even though it contains only a few per cent of the Galaxy's stars and contributes an even smaller proportion of its total light. The 'Population II' stars of the halo are generally so dim that it is difficult to spot them except when they draw attention to themselves. Some, for example, have reached the stage of life where they pulsate as variables of the RR Lyrae type. Halo stars that are currently passing through the Galaxy's plane near the Sun stand out by the fact that they seem to move quickly – though in fact, it is the Sun that is moving and leaving the 'high-velocity' stars behind.

A small fraction of the halo stars signpost their existence by banding together into distinct balls of stars – easily recognized through a telescope – called globular clusters. An obscure German astronomer called Abraham Ihle was the first to spot a globular cluster. As he followed the planet Saturn in 1665, he came across a fuzzy patch in Sagittarius, which was to be catalogued a century later as M22. In fact, there are two globular clusters that are much brighter than M22, and easily visible to the naked eye. But they lie so far south that northern hemisphere astronomers assumed that they were stars, and that their fuzzy appearance was caused by the atmospheric blurring of a star seen so low in the sky. The two brightest globular clusters are therefore named like stars, as omega Centauri and 47 Tucanae.

Omega Centauri was noted as long ago as the second century AD, by the Greek astronomer Ptolemy. Edmond Halley first realized it was a star cluster, on a voyage to St. Helena in 1677, and with a better telescope John Herschel later declared it 'truly astonishing . . . the richest and largest of its kind in

BOX 1. **Birth of the Milky Way**

By digging into the halo of the Milky Way, galactic archaeologists have found a wealth of data on the first stars to form in our Galaxy. And cosmologists are constantly finding new facts about the formation of galaxies in general. But, even so, there is no consensus as yet on the precise manner in which our Galaxy was born.

Cosmologists have proposed two competing ideas about the birth of galaxies. In the 'top down' model, the gases from the Big Bang first formed into very large structures – the precursors of clusters and superclusters of galaxies – which then broke down into smaller and smaller fragments that each condensed to form a galaxy. According to the 'bottom up' theory, the primordial gases clumped into comparatively small clouds, which were drawn together by gravity to form galaxies and clusters. Most researchers today support the 'bottom up' concept, but there are many variants of this theory that involve different sizes for the original clouds: each may have been as heavy as a galaxy, or just a fraction of this mass.

'Stellar archaeology' in our Galaxy has not, as yet, helped to settle this debate. In the 1950s, astronomers first measured the age of the halo and found that its stars were all much older than those in the disc. This prompted Olin Eggen, Donald Lynden-Bell and Allan Sandage to propose a simple explanation. In their classic 1962 paper – known as 'ELS' – these authors suggested that the Galaxy formed from a single cloud that was roughly spherical. The cloud was made almost entirely of hydrogen and helium left over from the Big Bang.

A small proportion of the gas in this cloud condensed rapidly into globular clusters and some isolated stars. This first generation of stars consisted of hydrogen and helium, with very little of the heavier elements that were created later inside stars and expelled into space. The surviving members of the first generation still continue to pursue orbits that carry them all around the same spherical region of space – the halo. But before most of the gas had time to condense into stars, it settled into the thin disc where most of the Galaxy's mass resides today. Here, heavy elements ejected from stars were incorporated into later generations.

But later observations have shown that the original theory – perhaps not surprisingly – is too simple. According to ELS, the gas in the halo collapsed in only a few hundred million years – a very short time on the cosmic scale – and so all the globular clusters should be much the same age. Moreover, the clusters further from the plane of the Galaxy should have formed first, and so contain the smallest amount of the heavier elements: these elements should be more common in globular clusters nearer to the plane of the Milky Way.

The first cracks in the ELS theory came in the late 1970s, when Leonard Searle and Robert Zinn in the United States found no sign of the predicted change in the abundance of heavy elements from the centre of the halo towards the centre. The simplest explanation was that the globular clusters in all parts of the halo covered a range of ages. This idea was dramatically supported a decade later, when new electronic detectors allowed independent teams from Canada and Italy to measure the ages of globular clusters with unprecedented precision. Even with the new measurements, the absolute ages of the clusters are a matter of some dispute, but they leave no room for doubt that the globular clusters were not born all at once.

These researchers have found, for example, that NGC 362 in Tucana is four billion years younger than the superficially similar NGC 288 in Sculptor, while the smaller Palomar 12 cluster, in Capricornus, is seven billion years younger than the ancient NGC 362. This range in age is ten times larger than is permitted by ELS.

Some theorists have argued that the new findings simply mean bringing in some refinements to the venerable ELS

FACING PAGE One of the first globular clusters to be identified, 47 Tucanae lies close to the Small Magellanic Cloud in the sky. It is easily visible to the unaided eye, appearing as a fuzzy 'star' of the fifth magnitude. At 13 000 light years away, 47 Tucanae's stars are about 10 billion years old – making this one of the younger globular clusters.

theory. Searle and Zinn suggested that some gas clouds remained in the outer part of the Galaxy's halo after the rest of the primordial gas collapsed to form the disc. It took several billion years for these residual gas clouds in the halo to condense into star clusters.

But others have advocated the total overthrow of ELS. In their version of the 'bottom up' theory, the basic building blocks were clouds of gas weighing only as much as a million Suns. The birth of the Milky Way – and other large galaxies – was not a single dramatic act, completed in only a few hundred million years. Instead, it was a prolonged and messy business, involving the amalgamation of millions of small clouds over a period of billions of years. Whenever a couple of gas clouds happened to collide at high speed, the shock would compress some of the gas into a cluster of stars – a newly born globular cluster. But most of the gas lost its energy by slower encounters, and gradually settled down into the disc.

Instead of forming in less than a billion years, this theory suggests that the Milky Way may not be complete even today: there are still small clouds of matter around our Galaxy, in the shape of the dwarf irregular galaxies. These small galaxies have undoubtedly condensed from intergalactic gas clouds in the comparatively recent past. They may still merge with our Galaxy to increase its mass a tiny jot more.

The new paradigm of galaxy-birth thus enhances the status of the dwarf irregular galaxies immensely. They are not simply cast-aside scraps, left over on the fringes of some great act of galaxy-creation. Instead, the dwarf galaxies are representatives of the basic stuff of the galactic Universe, the few survivors of the originally vast number of building blocks that went into creating our Galaxy and others.

Astronomers have found about 140 globular clusters in our Galaxy, and estimate that it contains about 200 in all, when we allow for the fact that some must be hidden behind the dense dust clouds in the direction of the Galaxy's centre. The tally is pretty well complete, however, because most of the globulars lie well away from the plane of the Milky Way. This distribution in the sky led Harlow Shapley – as early as 1918 – to propose that the globular clusters form a spherical swarm around the centre of the Galaxy. In his pioneering work, Shapley seriously overestimated the extent of this swarm, and therefore the size of the Galaxy as a whole. More recent measurements show that most of them lie in a region 100 000 light years across, defining the general extent of the Galactic halo.

But a few globular clusters – and some individual halo stars – lie even further out. By studying their motion, astronomers have been able to measure the total gravitational pull of the Galaxy, and so work out the total mass in the Milky Way. It turns out to be around a million million solar masses – about ten times the total amount of matter that we detect directly in the Galaxy, in the form of stars, gas and dust. So a surprising result emerges. Although the halo contains only a small fraction of the Galaxy's population of stars – and only tiny amounts of gas – it must contain 90 per cent of the Galaxy's total mass. And this mass must be in the form of some invisible and undetectable kind of matter.

Bizarre as it seems, the notion that most of the Galaxy's mass consists of unknown 'dark matter' is now generally accepted by astronomers. It ties in with measurements of the rotation of other galaxies, which show that most – and possibly all – spiral galaxies weigh about ten times more than the total mass of their stars and interstellar matter. Dark matter also makes its presence felt in clusters of galaxies. The individual members are moving so fast that the cluster would long since have split

the heavens, whose stars are literally innumerable'. This cluster is indeed not only the brightest in our skies, at magnitude 3.5, but the largest and most massive in the Galaxy. It is 600 light years across, and has a mass of just over a million Suns: because the average star in a globular cluster is less massive than the Sun, the number of stars in this one cluster must run into tens of millions – even if they are not literally 'innumerable'!

FACING PAGE The omega Centauri globular cluster in a different light – ultraviolet. This image of the largest globular cluster in our Galaxy was captured by ASTRO-1's Ultraviolet Imaging Telescope on a Shuttle mission in 1990. Many of the sources detected are invisible to ground-based telescopes, including 'horizontal branch' stars. These small hot stars are all that remains of red giants that were insufficiently massive to 'go supernova', but which have anyway ejected most of their outer envelopes.

up, unless the galaxies possess a far greater gravitational attraction that we would expect from the stars and other matter that we can detect. In fact, detailed theories of the Big Bang – in which the Universe began some 15 billion years ago – suggest that as much as 99 per cent of the Universe may consist of dark matter: the visible stars and galaxies may be just the conspicuous froth on the top of a deep dark ocean of invisible matter.

So far, astronomers have little idea of what the dark matter may be. The matter in the halo of our Galaxy – and others – may consist of something fairly ordinary, such as black holes or stars too low in mass to ignite – the so-called 'brown dwarfs'. In 1993, independent teams of astronomers found the first evidence to support this theory. Two groups monitored the brightness of millions of stars in the Large Magellanic Cloud, while a third monitored distant quasars. The brightness of these remote objects changed in the way predicted if a black hole or brown dwarf in the halo of our Galaxy moved in front and its gravity focused the light of the distant star or quasar.

But cosmologists believe the vast amounts of dark matter in the Universe as a whole cannot consist of ordinary matter, because nuclear reactions soon after the Big Bang would then have produced very different proportions of elements from those that exist in the Universe today. They suggest that the dark matter consists of so-far undiscovered subatomic particles, going by names such as axions or photinos.

These theories suggest that the dark matter was instrumental in sculpting the Galaxy. After the Big Bang, the dark matter clumped together to form invisible gravitational traps in space. Ordinary matter – at that time in the form of hydrogen and helium atoms – was then pulled into the gravitational wells, to condense into galaxies where the dark matter lay in the densest clumps. Regardless of the nature – or even the existence – of the dark matter, the earliest part of our Galaxy to form was undoubtedly the halo. Two strong pieces of evidence have convinced astronomers that the halo stars are the oldest in the Galaxy.

The first comes from the composition of these stars. The stars in the halo consist almost entirely of hydrogen and helium, with only very tiny proportions of the heavier elements that we find in the Sun and other stars that reside in the disc of the Galaxy. According to cosmologists, the Big Bang produced virtually nothing but hydrogen and helium. The other elements were forged inside stars, which then released these heavier elements into space in their death throes as red giants, planetary nebulae or supernovae. So we would expect the first stars in the Galaxy to consist almost entirely of hydrogen and helium – as we find in the halo stars – and successively later generations to contain more and more of the heavy elements, as is characteristic of the stars in the disc.

The second method involves a more quantitative measurement of the age of stars in the halo. Astronomers find it difficult to date a single star, but they can measure the age of a cluster of stars with more precision. The technique involves comparing the number of stars of different types in the cluster. When a cluster is born, the stars have properties that form a sequence: the more massive stars are progressively hotter and more luminous. On a diagram of luminosity versus temperature, they form a diagonal line called the 'main sequence'. The more massive stars age more quickly, turning into red giant stars, and then turning into completely different kinds of objects – or disappearing altogether – when they expire as supernovae or planetary nebulae. Because astronomers know the rate at which stars age, they can date a cluster by finding the most massive stars that are still on the main sequence.

Most clusters near to us – members of the Galaxy's disc – contain stars that are heavier than the Sun, and so are younger than our own star. But in the globular clusters, all the stars weigh in at less than 0.8 Suns. This means the globulars are around 12 billion years old – by far the oldest dated objects in the Galaxy. So the globular clusters must have formed well before any of the star clusters in the disc. The halo is, therefore, a fossilized relic dating from the earliest days of the Milky Way. As such, it provides our best insights into the way in which our Galaxy condensed from the swirls of gas left over from the Big Bang.

In fact, the halo is not completely devoid of gas even today – despite the 'textbook' view that the Galaxy's halo consists only of stars (and the enigmatic dark matter). In the 1960s, Jan Oort and his colleagues in Holland found patches of hydrogen gas well away from the plane of the Galaxy, and moving with high speeds – which are partly a reflection of the Sun's own motion around the Galaxy. No-one is quite sure what these 'high-velocity clouds' are. We do not even have any way to tell their distances, but most astronomers believe they lie in the Galaxy's halo.

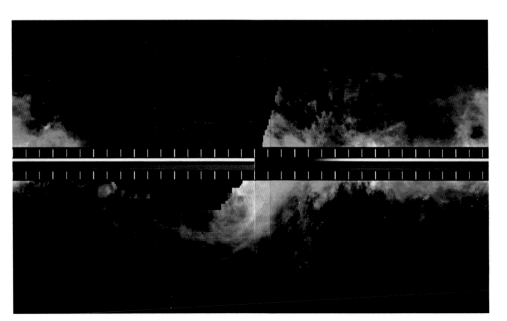

Living in the thick of our Galaxy, it is very difficult to work out its overall structure. An added complication is that space is 'foggy' with dust grains. Fortunately, dust grains are virtually transparent to radio waves, and the cool hydrogen gas in our Galaxy emits a signal at 21 centimetres. This velocity-coded map of interstellar hydrogen reveals that the gas curves in huge arcs over the sky (the black wedge is the area of sky that cannot be seen from the Hat Creek Radio Telescope in California). The colours are coded for velocity – red for recession, blue for approach, yellow for stationary gas – and demonstrate the large-scale rotation of our Galaxy.

According to one school of thought, the high-velocity clouds consist of gas from beyond the Milky Way. They may be streams of gas orbiting the Milky Way, like the Magellanic Stream that has been torn from our companion galaxies, or clouds of gas that have acquired their high speed as they fall into our Galaxy from far beyond. On the other hand, the high-velocity clouds may have been shot out into the halo from the gas-rich plane of our Milky Way, propelled by powerful supernova explosions. Individual clouds may be either travelling outwards with the momentum of the explosion or coming back to the plane of the Galaxy under the influence of its gravity, like a tossed ball falling back to Earth.

Turning from the invisible enigmas of the halo to the Galaxy's disc, we come to regions that are – in principle at least – much better understood. The disc and the central bulge consist mainly of stars, with a smattering of interstellar gas and dust. Roughly speaking, the disc is 100 000 light years across, and it is about 2000 light years thick – the proportions of two audio CDs placed one on top of the other. Towards the centre, the distribution of stars becomes a lot thicker, in a central bulge that is 6000 light years thick.

One of the problems of living within our Galaxy is that we cannot tell the shape of the central bulge. Most textbook diagrams assume – for simplicity – that if we could look on our Galaxy from above, the bulge would be circular in shape. But we know that the central region of at least half of all spiral galaxies is in the form of a straight 'bar', with the spiral arms springing from the end of the bar. The thinnest bars are five times as long as they are wide, and the proportions range from this extreme to galaxies where the central bulge is circular.

Seen sideways-on, a central bar would not look much different from a simple circular bulge. But there are two reasons for thinking that the central bulge of our Milky Way is bar-shaped, to some extent at least. First, the Infrared Astronomical Satellite (IRAS) picked out several thousand individual stars in the galactic bulge, and the stars to one side of the galactic centre are on average slightly closer to us than the stars on the other side – suggesting that we are seeing a bar of stars inclined to our line of sight. Second, there are fast-moving clouds of gas about 5000 light years from the galaxy's centre whose motion is most easily explained by the gravitational influence of a bar as it rotates. It's most likely, then, that the central bulge of the Milky Way is slightly bar-shaped – perhaps twice as long as it is wide, as we see, for example, in the spiral galaxies M61 and M83.

For many years, astronomers thought that the Milky Way's

The Milky Way's twin? This CCD image of the spiral galaxy M61 in Virgo shows a galaxy that could resemble ours. In particular, its central regions are not spherical – it has a barred distribution of matter that may be similar to the distribution of stars at the centre of our own Galaxy. Additionally, its spiral arms have unusually sharp bends in them. One of the problems in mapping our Galaxy has been the assumption that its spiral arms are smooth, curving and connected. If – instead – we allow for bends and discontinuities, there seems to be a better fit.

bulge was simply the dense inner region of the halo – made of the same kind of stars, packed more closely together. But in the 1960s, the American astronomer Halton 'Chip' Arp found that although these stars were definitely very old, they contained much more of the heavy elements than the hydrogen–helium stars of the halo. This means that several generations of stars must have lived and died in the bulge before the present stars were formed. And IRAS found that the bulge contains thousands of giant stars that are far brighter than any stars of the halo.

Instead, the bulge may be related to the stars of the 'thick disc' – a region that stretches for several thousand light years above and below the visible disc of stars. For several years, New Zealand astronomer Gerry Gilmore has argued that this region contains a population of stars intermediate between those of the halo and the disc. These stars are embedded in gas that is very tenuous and extremely hot, at a temperature

of 100 000 degrees Celsius. Astronomers have been able to study this gas only because it absorbs ultraviolet radiation from distant objects, such as the supernova that exploded in the Large Magellanic Cloud in 1987. The faint stars and the tenuous gas of this region are difficult to study, however, and the thick disc is, as yet, little explored.

If we dissect a slice through the Galaxy's disc, from the top to the central plane, we find that its 'Population I' stars show a sequence of layers, consisting of objects of different ages – rather like geological strata. The topmost layer is the – still-controversial – thick disc, with its elderly stars. Next comes the main disc, a couple of thousand light years thick, which contains the vast majority of the Galaxy's stars, including the Sun. The thinnest layer, lying right in the plane of the Milky Way and only 500 light years thick, contains the bulk of the Galaxy's gas and dust, and the very youngest stars (extreme Population I) that are freshly born from this interstellar material.

What the layer of gas lacks in thickness, however, it makes up in width. It stretches outwards to a distance of 80 000 light years from the galactic centre, while the stars of the Milky Way peter out at a radius of 50 000 light years. The outermost parts of this thin gas disc are warped like the brim of a hat. On one side – in the direction of Cygnus – the layer of gas tilts upwards until it reaches a height of 10 000 light years above the galactic plane. On the other side, the gas layer tips below the plane of the Galaxy, but bends back upwards again further out.

Our Galaxy is far from unique in its hat-brim structure: about 50 per cent of spiral galaxies have a warped disc of gas, and in some cases it is even more pronounced than the twist in the Milky Way's outer regions. The gas could be warped by a gravitational tug from a companion galaxy – the Large Magellanic Cloud in the case of the Milky Way. But the calculations do not fit well, and some highly warped galaxies do not have companions anyway. In 1984, British astronomer Linda Sparke suggested that the warps are caused instead by the massive amounts of dark matter in the halo. This dark matter should have a strong gravitational pull on the outermost regions of the gas disc. If the distribution of dark matter is slightly skewed, relative to the visible galaxy at the centre, then it would bend the outer edge of the thin gas disc in just the way that we observe.

The Galaxy's disc is the only region that contains a signifi-

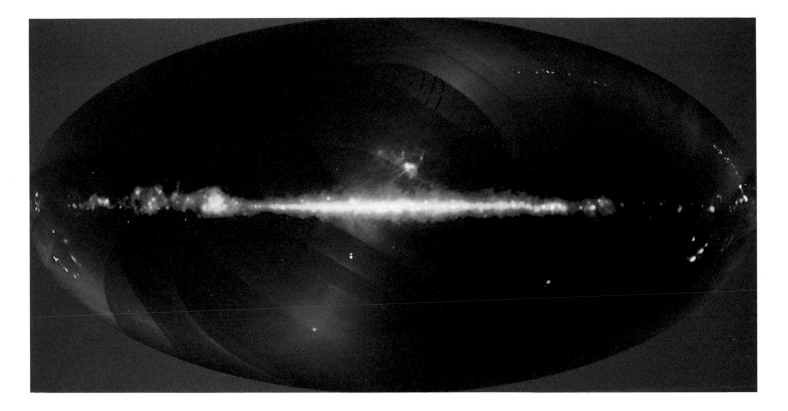

View of the whole sky, obtained at far-infrared wavelengths (25, 60 and 100 micrometres) by the Diffuse Infrared Background Experiment on COBE. Infrared radiation penetrates the interstellar dust and helps astronomers to unravel the structure of our Galaxy. The bright complex on the left is the Galaxy's Cygnus Arm, seen end-on. (The faint blue S-shaped band running from top to bottom comes from dust within our Solar System.)

cant amount of material between the stars. As well the obvious gas and dust, the interstellar medium contains several less-substantial components: these do not have any significant mass, but do carry energy with them.

The first is so obvious that it is easy to overlook: the energy of photons of radiation travelling through space. Most of this radiation is in the form of light and ultraviolet from the bright young stars in the disc. The ultraviolet plays an important role in splitting up atoms in space. More surprisingly, one-third of the total radiation energy in the disc consists of radio waves left over from the Big Bang – part of the 'microwave background' that fills the whole of the Universe.

The second component is also obvious to anyone who glances even casually at the Milky Way – although its true nature was not known until this century. The bright background of stars is broken in places by dark 'rifts'. The most famous is the Coal Sack, next to the Southern Cross, but an even larger rift stretches through the northern constellation of Cygnus. Throughout history, astronomers have thought these were genuine gaps in the distribution of stars, and their existence had a major influence on the models of the Milky Way drawn up by William Herschel. But the great American pioneer of astrophotography, E. E. Barnard, found many more dark patches. He noted in 1919 that these patches looked like dark clouds silhouetted against the bright background, and absorbing the light from stars behind.

Barnard's idea of absorbing matter in space was confirmed by the Swiss astronomer Robert Trumpler. Working at the Lick Observatory in California, Trumpler found that distant stars were both dimmed and reddened in colour as their light travelled through interstellar space. On average, every 3000 light years of space robs a star of half its light, dimming it by almost a magnitude. Trumpler proposed that this was caused by fine 'dust' spread thinly between the stars: where the dust is more concentrated, we find Barnard's dark clouds and the larger rifts.

From the way that the grains affect starlight, astronomers have deduced that they are tiny chips of silicate – rather similar to the Earth's 'rocks'. The grains are typically half of a micrometre across, so it would take about 200 laid side by side to stretch across the width of a human hair. Many of the grains are coated in a layer of organic material, created by chemical reactions triggered by ultraviolet radiation. In the densest clouds, the grains have a frosting of ice. These silicate grains were expelled from old red giant stars that were rich in silicon and oxygen.

One red giant in ten is rich in carbon, instead. These stars eject particles of 'soot' that are much smaller than the silicate grains, only 0.01 micrometres in size. The grains may literally be made of soot – amorphous carbon, which would appear as just a jumble of carbon atoms if we could magnify it sufficiently – or soot with hydrogen atoms stuck all over the surface (hydrogenated amorphous carbon). They may also contain a form of carbon only recently found on Earth: a spherical molecule of 60 carbon atoms enclosing a hollow space. The molecule looks like a football, or like two geodesic domes stuck back to back: it is officially named buckminsterfullerene, after the scientist and architect Buckminster Fuller, who invented the geodesic dome, but most scientists call it a 'buckyball'.

The dust in space led astronomers to another of the Galaxy's less substantial components, its magnetic field. The first hints of this field came in 1949, when John Hall at the US Naval Observatory and Albert Hiltner at the McDonald Observatory in Texas found that light from stars is slightly polarized. The amount of polarization was related to the amount of dust between the star and us. Hall and Hiltner realized that the magnetic field was lining up the elongated grains of dust, so they formed a screen that polarized light coming through them. Later, radio astronomers found that the magnetism in the Galaxy's disc also affects the polarization of radio waves from pulsars and other distant objects.

The magnetic field has a strength about one-millionth that of the Earth's magnetism. No-one is quite sure how the Galaxy's magnetism came about. The original field may have been tied into the clouds of gas from which the Milky Way was born. Or the magnetism may be generated by the Galaxy's ionized gases as the Milky Way rotates, rather like a cosmic electromagnet. Either way, we only need to produce a very small original field. The magnetic field is firmly tied into the interstellar gas, and this gas is continuously kept well stirred by the explosions of supernovae and the effects of the spiral arms. These increase the energy of the magnetic field, in much the same way as you can increase the energy stored in a rubber band by twisting it.

Astronomers had in fact stumbled across signs of the Galaxy's magnetism in the 1930s, without realizing it. In 1931, a young electrical engineer called Karl Jansky began to check out sources of radio interference with an antenna at Holmdel, New Jersey. He found that some of the annoying 'hiss' came from the sky, and suggested that the radiation emanated from the band of the Milky Way. Twenty years were to pass before astronomers found out why our Galaxy is a powerful natural broadcaster. The answer involved both the widespread magnetism within the Galaxy and another component of the interstellar medium, discovered even earlier than the radio emission: cosmic rays.

From the turn of the century, physicists had known that there is a background of penetrating radiation all around us. They had assumed that it came from radioactive elements in the Earth's crust. On 7 August 1912, an Austrian physicist called Viktor Hess set out to test this idea. He took a radiation detector up in a balloon, expecting the amount of radiation to fall off with increased distance from the Earth. Instead, the intensity of radiation increased as he went higher. Hess concluded 'the results of these observations seem best explained by a radiation of great penetrating power entering our atmosphere from above'.

These cosmic rays, we now know, consist of high-speed electrically charged particles. Most of them are protons, with a mix of heavier atomic nuclei and a small proportion of electrons. Some gigantic natural 'particle accelerators' must be responsible for these energetic particles. Supernova remnants

– the expanding shells from old star explosions – are probably responsible for the bulk of the cosmic rays, those with the lowest energies. The higher-energy cosmic rays probably come from unusual double stars called X-ray binaries. Here, gas from an ordinary star is dragged away by a very compact neutron star. The gas swirls around the neutron star in a hot disc, where individual particles can pick up enough energy to shoot away at high speeds. One X-ray binary on our side of the Galaxy, called Cygnus X-3, may on its own be providing several per cent of the Galaxy's high-energy cosmic rays. The fastest cosmic rays of all seem to come from beyond our Galaxy, and have probably travelled to us from the exploding centres of other galaxies.

Most of the cosmic rays are trapped inside our Galaxy by its magnetic field. The lightest of the cosmic-ray particles, the electrons, feel this force the most. As they whirl around in the interstellar magnetic field, the electrons produce radio waves by the synchrotron process – first observed on Earth in the particle accelerators called synchrotrons. This is the natural 'radio interference' that Jansky first detected in the 1930s.

In the decades after Jansky, radio astronomy came to have the key role in unravelling the structure of the Galaxy. The basis was a remarkable prediction made during the Second World War in occupied Holland. Unable to use any telescopes, Dutch astronomers turned to theoretical problems. The radio waves detected so far from the Milky Way covered a whole continuum of wavelengths, and Jan Oort – a pioneer of the study of the Milky Way – asked his students if any gases in space would emit radio waves at just a single wavelength. Henrik van de Hulst came up with an answer. Hydrogen atoms should emit at a wavelength close to 21 centimetres. In 1951, three groups of radio astronomers around the world picked up this characteristic radio signal from hydrogen between the stars.

Strictly speaking, this was not the first detection of gas in space. In the early years of the twentieth century, Johannes Hartmann at the Potsdam Astrophysical Observatory was investigating the spectra of stars in Orion's Belt when he found narrow absorption lines due to calcium atoms in the space between the stars and us. But such measurements can be made only where there is a star fairly near to us: the hydrogen measurements opened up the entire Galaxy.

Hydrogen is indeed the most common element in the inter-

Viktor Hess, an Austrian-American physicist, performed a series of balloon flights to locate the source of 'penetrating radiation' that was known to be all around us. To his surprise, he proved that the radiation came from space. It was later found that it was 'cosmic radiation' – high-speed electrically charged particles that race around the Galaxy. For his discovery of cosmic rays, Hess was awarded the Nobel Prize for Physics in 1936.

Although cosmic rays come from the depths of space, they are still capable of packing a punch when they arrive on Earth. In this false-colour image, a cosmic-ray sulphur nucleus (red) has just collided with a nucleus in the photographic emulsion. The collision has produced a spray of other particles: a fluorine nucleus (green); other nuclear fragments (blue); and 16 pions (yellow).

Although Cygnus X-3 appears only as a bright blob in this image from the Rosat X-ray telescope, and is invisible in optical telescopes, it is one of the most powerful objects in our Galaxy. Lying 25 000 light years from us, Cygnus X-3 consists of a compact neutron star pulling gas from a companion, and whirling some of the particles almost up to the speed of light. This object alone produces a significant amount of the Galaxy's high-energy particles, the cosmic rays.

stellar gas, making up nine atoms out of ten. Virtually all the rest is in the form of helium. These are the two gases that have survived since the Big Bang, and they are joined by a smattering of heavier elements that have been created in stars and then jettisoned into interstellar space. The number of heavyweight atoms floating freely in space is roughly equal to the number bound up in solid grains of interstellar dust: even taken together, however, these heavyweight atoms are outnumbered a thousand to one by the hydrogen.

Because the interstellar gas is invisible to our eyes, it is easy to think that it is spread out uniformly, like the gas in the Earth's atmosphere. But the exact opposite is the case. If we travelled a thousand light years through the interstellar medium, we would traverse regions with greater differences than the contrast between the Earth's air and our planet's oceans: from stretches of tenuous hot gas at a temperature of a million degrees to small clouds that are a million times denser and have a temperature just a few degrees above absolute zero.

Astronomers have identified four main components of the interstellar medium. Most of the matter is spread out in the 'warm intercloud medium' – where 'warm' means a temperature of 8000 degrees Celsius, or somewhat hotter than the surface of the Sun! Interspersed in this warm medium are bubbles of the million-degree tenuous gas, blown by stars that have exploded as supernovae.

The third component consists of dense and cold clouds of hydrogen atoms. In the 1970s, American astronomer Carl Heiles found that these clouds are not round and fluffy, but long and thin – resembling cirrus clouds in our atmosphere. Many of these filaments form long curved arcs, delineating the edges of large shells of gas. These huge shells have been inflated either by the explosion of many ancient supernovae or by the hot gases from clusters of young stars. The dust grains mixed with this gas are unusually good at emitting infrared radiation, and so these filaments showed up unexpectedly in the pioneering survey of the sky by the Infrared Astronomical Satellite in 1983 as 'infrared cirrus' covering most of the sky.

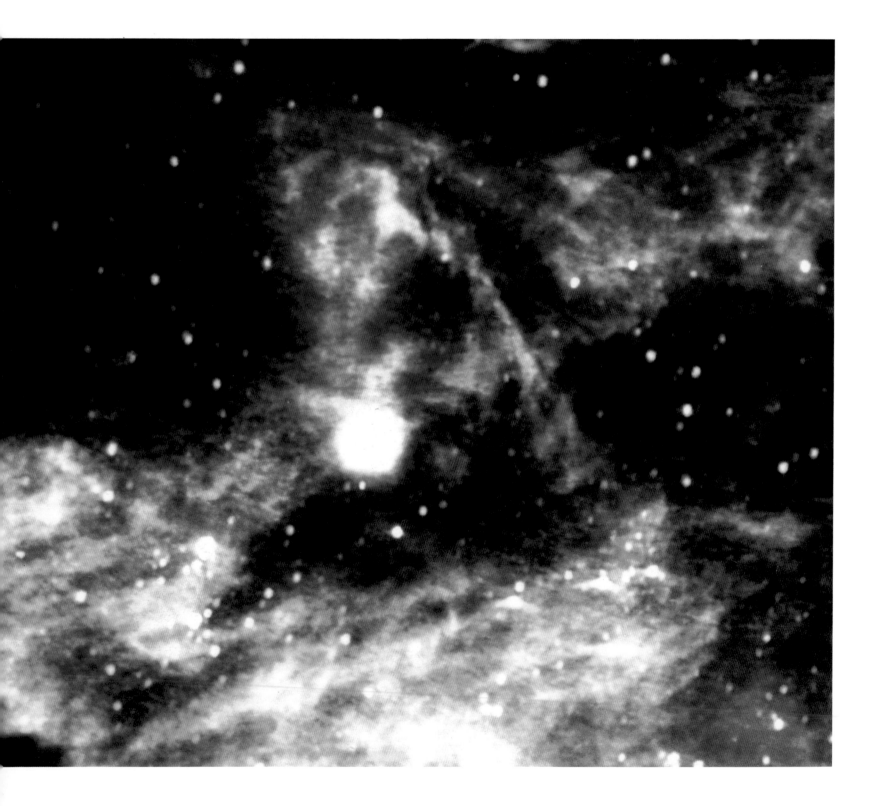

Astronomers are still not quite certain how these first three components of the interstellar medium are related. At one extreme are astronomers who believe that the hot bubbles are relatively small, and take up in total only 10 to 20 per cent of interstellar space in the Galaxy's disc: the rest consists of the warm gas, laced by the denser threads of gas in the cold hydrogen clouds. The diametrically opposed view is that the hot bubbles fill most of interstellar space – like an exceptionally 'holey' Swiss cheese – with the dense clouds forming their boundaries, and the warm medium being rather unimportant. Most of the bubbles would then connect with one another. Many bubbles would also extend up and down from the plane of the Galaxy to such an extend that their hot contents would erupt into the much emptier region of the halo. In the 'fountain model', this hot gas would then cool and fall back to the galactic disc in the form of high-velocity hydrogen clouds.

The fourth main component of the interstellar medium consists of relatively small but very dense clouds. Ironically, these have been known for longer than the more tenuous phases, because the absorbing dust in these clouds makes the nearer ones appear in silhouette against the background Milky Way, as the dark patches and rifts that Barnard recognized to be 'dark nebulae'. Astronomers soon came to realize that these dark clouds are the birthplaces of new stars, but the dense dust totally hides the scene of stellar nativity from optical telescopes. Oddly enough, studies of the hydrogen atoms in these clouds did not help much: their 21-centimetre radiation was surprisingly weak.

A twin breakthrough came in the 1960s, with two technologies that can see through the dust. The first infrared telescopes revealed young stars hidden in dense dark clouds, including a star over 10 000 times brighter than the Sun that lurks deep in a dark cloud immediately behind the Orion Nebula. More important, astronomers picked up radio waves from molecules in these clouds.

As with the discovery of atoms in space, molecules first made themselves known to optical astronomers, by the absorption lines they imprint in the spectra of stars. Around 1940, Walter Adams at the Mount Wilson Observatory, California, discovered compounds of carbon with hydrogen (CH) and nitrogen (CN). Most simple molecules broadcast radio waves at certain characteristic wavelengths, and in the 1960s radio

FACING PAGE Part of the interstellar medium consists of cold, dense clouds of hydrogen atoms. Radio astronomers had for long suspected that these clouds were filamentary and thin – and this was confirmed by the IRAS infrared satellite in 1983. IRAS detected 'lukewarm' radiation from the dust grains mixed up in this gas. These form an 'infrared cirrus' – seen here in false colour in the constellation of Chamaeleon – which covers the whole sky.

Working at Mount Wilson Observatory in the late 1930s and early 1940s, Walter Adams was the first to detect molecules in interstellar space. Examining distant stars, he discovered molecules such as cyanogen (CN) and methylidyne (CH) in their spectra, arising in the cool gas and dust along the line of sight.

astronomers began to tune into these signals. They found hydroxyl (OH), water (H_2O) and formaldehyde (H_2CO). By the early 1990s, astronomers had detected about 80 different molecules in space – including some that contain ten or more atoms.

Most of these molecules are found in the dense clouds – now generally known as 'molecular clouds' – where the dust shields them from the disruptive effect of the general ultra-violet radiation in the Galaxy. Astronomers quickly realized that in these dense clouds, individual atoms of hydrogen – the commonest element – must also combine in pairs into hydrogen molecules (H_2). These molecules do not emit the 21-centimetre radiation: indeed, they emit very little radiation of any kind.

Nonetheless, it is possible to trace the distribution, temperature and motion of hydrogen in a molecular cloud by studying another molecule that is well mixed in with the hydrogen. Astronomers have chosen carbon monoxide as their tracer. Typically, there is only one of these molecules to every 10 000 molecules of hydrogen, but carbon monoxide is easy to detect because it emits powerfully, at a wavelength of 2.6 millimetres.

An average molecular cloud, it turns out, contains as much matter as half a million Suns, and is 120 light years across. But there is a wide range. The famous Coal Sack is one of the smallest molecular clouds we know, weighing in at only 4000 Suns, and is prominent only because it lies near to the Sun. At the other end of the scale, the Galaxy contains a couple of dozen 'giant molecular clouds' that contain almost 10 million solar masses of gas. Taken together, the molecular clouds contain about 2500 million solar masses. This is almost as much as all the other components of the interstellar medium, and the discovery of molecular clouds in the 1970s meant that astounded astronomers had to double their estimates of the total amount of gas in the Galaxy.

Molecular clouds are the maternity wards for newly born stars. The birth of a star is heralded by a denser 'core' in the cloud, that shows up in the spectral line of a molecule like ammonia. As gravity draws this material together, it heats up and begins to shine at infrared wavelengths, as a protostar. Nuclear fires then ignite at its centre, and a star is born. The moment of star-birth is however difficult to measure, because the heat generated by nuclear reactions is at first considerably less than the energy liberated by the gas that is still falling in.

Astronomers using the orbiting Infrared Astronomical Satellite and ground-based infrared telescopes have found many cases where a dense core contains an infrared source. Most of these are stars that have already begun to shine. Only in the 1990s did astronomers begin to unearth the 'Holy Grail' – infrared sources that are almost certainly genuine protostars.

The powerful combination of infrared and the radio study of molecules has shown that the smaller molecular clouds only produce stars that are relatively light in weight – not much

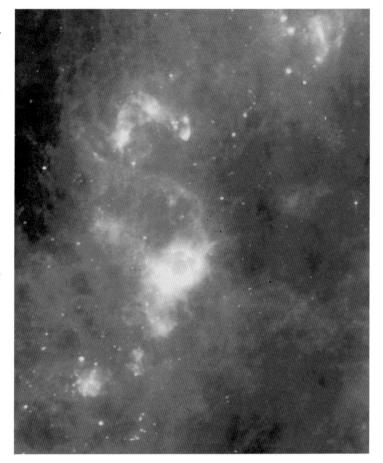

The IRAS infrared satellite discovered a great many places in the Galaxy where star-birth is taking place in heavily obscured regions. This image of the region around the Rosette Nebula in Monoceros not only shows the Rosette (centre), but many other places where stars are being born – but which are not visible in optical telescopes. The small blue dots are red giant stars.

heavier than our Sun. The giant molecular clouds are responsible for spawning the most massive stars, up to a hundred Suns; they produce relatively few of the lightweight stars.

The course of a star's life is determined almost entirely by the mass it is born with (the other major influence is whether it has a companion, because a pair of stars can swap mass as they age). To begin with, all stars shine by a nuclear reaction that converts hydrogen in their centres to helium – like a hydrogen bomb going off in slow motion. Because they are powered by the same basic reaction, these stars form a regular progression – the 'main sequence' in which the more massive stars are progressively brighter and hotter. Although the more massive stars have larger reserves of fuel, they burn so fiercely that they get through the central hydrogen in only 10 million years. A mediumweight star, like the Sun, continues to burn hydrogen for 10 billion years.

After the main sequence phase, the core of the star begins to shrink. At the same time, its outer layers billow out to about a hundred times its previous size and cool to a dull red glow. The star has become a red giant. Many red giants pulsate in size, changing in brightness as they swell and shrink. The most regular pulsators are the Cepheid variables, which can be as bright as 100 000 Suns.

An old star also sheds much of its outer gases to space. The stars that started out as the heaviest are the most desperate to lose weight in their final years. The Swiss astronomer André Maeder has calculated just what the effect will be, after these massive stars have 'evaporated copiously'. In 1990, he concluded that a star born 30 times heavier than the Sun will lose two-thirds of its matter in hot streams of gas – a hurricane version of the solar wind that the Sun currently emits.

This analysis explains one of the strangest types of star in our Galaxy: the Wolf–Rayet stars. As early as 1867, the French astronomers C. Wolf and G. Rayet discovered some stars with an unusual spectrum: instead of dark absorption lines, they had bright emission lines. In some Wolf–Rayet stars, these lines come from nitrogen; in others, from carbon and oxygen. More recently, astronomers have found that these stars are extremely hot (with surfaces around 50 000 degrees Celsius) and among the most luminous known, some 100 000 times brighter than the Sun. Astronomers can explain all these unusual properties if Wolf–Rayet stars are, in Maeder's words, 'bare cores left over by the peeling of massive stars'.

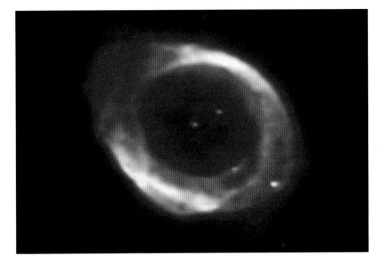

A CCD image of the planetary nebula M57 shows a star on the way out. Taken in the light of the red line of hydrogen at 656 nanometres, the image reveals the hot core of the ageing star – probably once very similar to our Sun – surrounded by the envelope of gases it has puffed off. The shell is expanding at about 20 kilometres per second and will disperse into space within a few thousand years. Meanwhile, the core – now a white dwarf star with no source of nuclear energy – will slowly leak away its heat into space.

The final fate of a star depends critically on its mass. If a star began life up to eight times heavier than the Sun, it shrugs off most of its remaining layers as a planetary nebula – a shell of gas that shines because it is lit up by ultraviolet radiation from the shrunken core of the original star. When the nuclear reactions stop altogether, this core becomes a white dwarf: a ball of matter that is a million times denser than water. It may consist of helium, or iron or an oxygen–carbon mixture, but it is the pressure of electrons that keeps its matter from collapsing yet further under its own powerful gravity.

In the core of a heavyweight star, nuclear reactions build up elements as heavy as iron, and the pressure tries to engender still more reactions. But the creation of heavier elements takes in energy, rather than releasing it. As a result, the flow of energy from the centre stops, and the core collapses. A sudden surge of energy – in the form of shock waves and neutrinos from the superhot core – then rips away the star's outer layers, in a supernova explosion.

The massive star's core continues to collapse. It may stop shrinking when it is 25 kilometres across, and electrons and protons have been pushed together to make an object consisting solely of neutrons. Such a neutron star is so dense that a pinhead of its material would weigh a million tonnes. Young neutron stars are often magnetized and emit radio waves: they are known as pulsars if their rapid spin makes the radio emission appear to flash like a light-house beacon.

If the core is more massive than three Suns, however, even the force between neighbouring neutrons cannot stop its collapse. It shrinks to – in theory – zero size and infinite density. Even if conditions do not become quite that extreme, the object acquires such a strong gravitational field that nothing – not even light – can escape. It is a black hole.

Star-death is not the end of the story. It is only part of a continuing cycle, for a star is a phoenix: from its ashes rises a new generation of stars. Red giant stars, planetary nebulae and supernovae all eject matter into the interstellar medium. And it is not just simple recycling, for this matter is enriched in new elements made inside the stars: this is how the composition of the Galaxy has changed from the original hydrogen–helium mixture created in the Big Bang. In particular, the heavy elements forged in stars are responsible for the dust in space, which has led – among other consequences – to the existence of solid planets like the Earth.

In addition, the shock waves from supernovae are important in sculpting the interstellar medium, squeezing the tenuous gas into denser clouds, where new stars may be born. The Dove Nebula in Canis Major, for example, is the shell of an old supernova that now carries a clutch of newborn stars on its wings. And so the cycle is completed, with supernovae seeding the interstellar medium with new elements, and compressing its gases into molecular clouds where new stars are born.

Supernovae also play a role in determining the overall 'personality' of our Galaxy. One of the attractive features of spiral galaxies – as opposed to the bland ellipticals – is that each has an easily recognizable and distinct shape. An extragalactic astronomer can instantly put a name to dozens of the brightest spiral galaxies, on the basis of their appearance alone.

In 1978, Philip Seiden and Humberto Gerola of the IBM Research Laboratory ran a computer simulation of the effect of a chain of supernova explosions, each squeezing the surrounding gas and creating a batch of new stars where the heaviest stars, in turn, exploded as supernovae. The sequence of supernovae squeezed the interstellar gas and dust into several elongated regions of denser material. As the galaxy's rotation twisted these regions around, they formed into sections of spiral arm before eventually dissolving.

This idea challenged the existing theory of spiral arms, which had been put forward by Chia-Chiao Lin and Frank Shu in the 1960s. The density-wave theory starts with a galaxy acquiring a spiral shape by some accident – perhaps because it is stirred up by the gravity of a passing galaxy, like cream in a cup of coffee assuming a spiral shape when you stir it. Lin and Shu showed that this spiral pattern would persist more or less for ever, even though individual stars and gas clouds are always drifting into the arms and out again. The spiral is a density wave that keeps its shape as its actual constituents change. In many ways, it is like a traffic jam on a motorway, which carries on long after the original obstruction has been cleared; the vehicles making up the jam change continuously as cars and lorries leave at the front and others join at the rear.

In real galaxies, both these processes are working. When we see a beautifully elegant two-armed spiral, extending for two or more turns around the galaxy's centre, we are encountering the work of a galaxy-wide density wave. One grand example is M100, in Coma Berenices, which is the largest spiral in the Virgo Cluster of galaxies. But these 'grand design spirals' – in the terminology of American researchers Debra and Bruce Elmegreen – account for only one in ten spirals. Most are a lot tattier in appearance. The opposite extreme, 'flocculent spirals', have no overall structure, but are composed of masses of short spiral segments. These look like the work of chains of supernovae. The Elmegreens select M63, in Canes Venatici, as the prototype flocculent spiral. Even an amateur's telescope will show that the 'arms' of M63 appear only as 'a mottled haze', in the words of American observer David Eicher.

Most galaxies lie between the two extremes. There is often

FACING PAGE This infrared image of the Cygnus Loop, made by the IRAS satellite, shows what remains of a star that exploded as a supernova some 20 000 years ago. The Loop, coded yellow-green in this image, is a shell of dust being heated by the shock wave from the old supernova. Behind it (coded red) are the more distant infrared cirrus clouds in the interstellar medium.

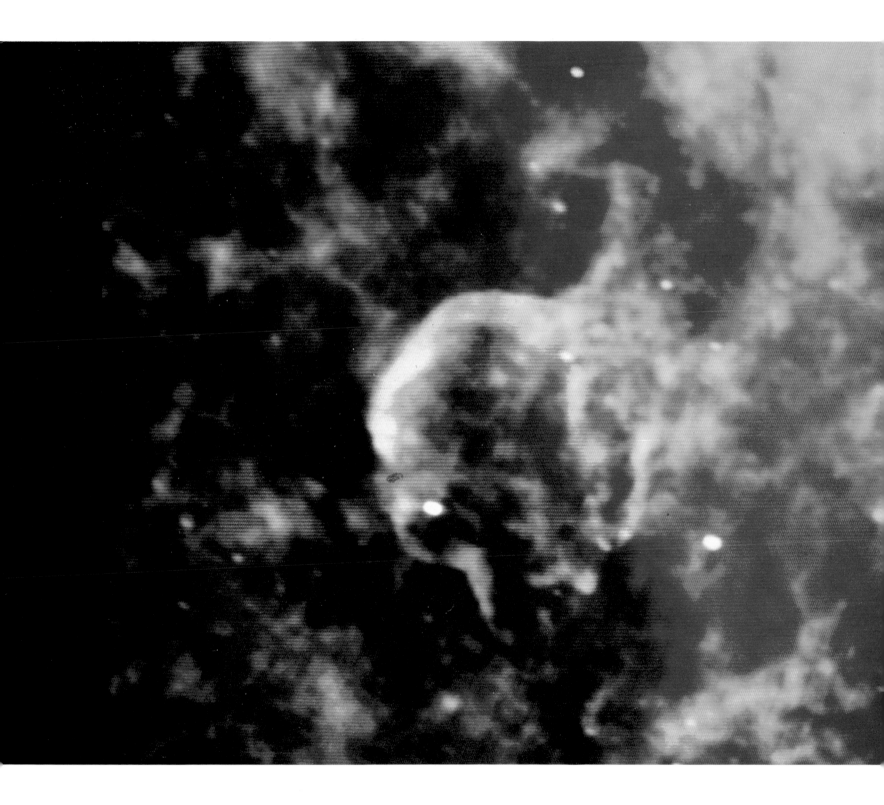

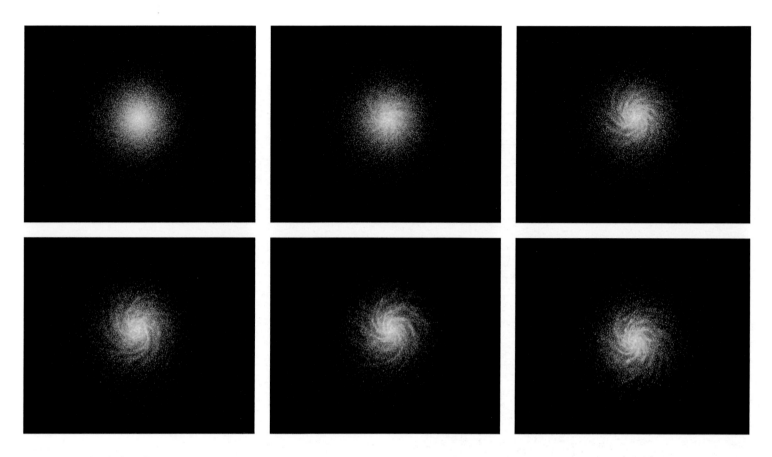

A computer simulation shows just how easy it is to make spiral arms in a galaxy, according to the density-wave theory. This sequence of images depicts a collection of dots – representing stars – moving as the galaxy rotates. Once a spiral pattern has formed, its gravity makes the shape persist, even though individual stars are constantly moving into the arm and out again.

FACING PAGE Spiral galaxies are often pictured as looking like M100 (top left), with bold, continuous spiral arms. But these 'grand design' spirals account for only one in ten of all spiral galaxies. Many have structure that is far less well-defined, with galaxies like M63 (bottom left) being the other extreme. These 'flocculent' spirals have spiral arms that resemble flocculent clouds: the structure is disconnected, and the arms are made of spiral segments rather than sweeping arcs. The difference between the two types shows up even more dramatically on the computer-enhanced images of the same galaxies rotated so they look flat-on (right). Detailed analysis of images like these will enable astronomers to find out how these different structures originate.

a distinct spiral pattern near the centre, but a tangle of smaller segments further out – sometimes giving the appearance of four or more arms – where the localized effect of supernovae overpowers the overall gravitational pattern of the density waves.

Grand designs and flocculent patterns – and all the intermediates – occur in all the Hubble types of spiral galaxies, from the tightly wound spiral arms of the Sa galaxies, through Sb galaxies like Andromeda to the open-armed Sc type exemplified by the Triangulum Galaxy M33. The extended classification adds to the decades-long debate about the 'personality' of our Milky Way: what is the characteristic shape of our Galaxy, hidden from us because we lie within this star-system, but well-known to any inhabitant of the Andromeda Galaxy?

We can make a lot of headway towards finding the Hubble type of the Milky Way without even knowing how the spiral arms would look from outside. Edwin Hubble himself found that Sa galaxies have comparatively large bulges in the centre, whereas Sc galaxies have small bulges. Even from our location in the Milky Way we can measure the size and brightness of our Galaxy's bulge, especially with data from infrared satellites that can see through the dust that lies in the way. This rules out the relatively fat Sa galaxies and almost-bulgeless Sc types. Most likely our Galaxy is either Sb or Sbc – intermediate between Sb and Sc.

In the 1970s, Gerard de Vaucouleurs of the University of

The Diffuse Infrared Background Experiment aboard the COBE satellite captured this view of the Milky Way at near-infrared wavelengths of 1.2, 2.2 and 3.4 micrometres. It dramatically reveals the Galaxy's spiral nature (seen edge-on), along with the slightly flattened nuclear bulge. It also allows astronomers to classify the morphological type of the Galaxy from its disc/bulge ratio. It comes out as intermediate between types Sb and Sc – as an Sbc galaxy.

Texas at Austin concluded that the complete Hubble type of the Milky Way is 'SAB(rs)bc II'. The 'SAB' sums up the idea that our Galaxy has a weak central bar, intermediate between that of galaxies with a round central bulge (SA) and those (SB) with a very prominent bar. De Vaucouleurs's '(rs)' class means there is a weak central ring of stars and gas: type (r) have definite rings whereas (s) have no rings. The basic Hubble type is 'bc', whereas the final II refers to the Galaxy's overall luminosity: it is in the second league here, as compared with the much brighter Andromeda Galaxy, which rates a 'I'.

Other astronomers have criticized de Vaucouleurs's classification: plumping for intermediate properties in every way can look suspiciously like being uncertain of anything! But subsequent research has by and large backed up de Vaucouleurs's conclusion: American astronomer Virginia Trimble catagorizes our Galaxy as 'disgustingly normal'. According to de Vaucouleurs, M61 in the Virgo Cluster of galaxies is the Milky Way's best lookalike, showing how the short bar, partial ring and inner spiral arms would appear from outside.

When it comes to tracing the arms of our Galaxy, astronomers are in the position of a terrestrial surveyor deprived of aerial photographs. They can only map our Galaxy by painstakingly measuring the distances to stars, nebulae and gas clouds, and plotting their positions. If we want to map the spiral arms, there is no point in just plotting all the stars we see. For a start, most of them lie right on our doorstep on the scale of the whole Galaxy. In addition, the spiral arms are not especially thick with ordinary stars.

The spiral pattern is part of the galaxy's disc, and simply marks regions where matter is more concentrated. The actual number of stars per cubic light year in a spiral arm is only ten per cent higher than in the rest of the disc. When we photograph other galaxies, the spiral arms are prominent because the interstellar matter is bunched up here, and is frenetically forming into new stars: in the words of the pioneering galactic astronomer Walter Baade 'the arms of the Andromeda Galaxy are much like the candles and frosting on a birthday cake – all show and little substance'. A galaxy's arms contain relatively little of the mass of the galactic disc, but they do have extremely brilliant massive stars – of types O and B – that are short-lived and do not travel far from their birth places before they explode as supernovae. The arms are also strung with bright glowing nebulae that are lit up by the young stars.

BOX 2. **Spectral types and the H–R diagram**

With the naked eye, we can see that the brightest stars are very slightly different colours: Orion sports the two extremes, with red Betelgeuse and blue-white Rigel. The colours result from the differing temperature of the stars. By studying star temperatures in detail, astronomers have derived a method of classifying stars that provides a deep insight into their fundamental nature.

With electronic detectors, astronomers can find the colour of a star with great precision. It is just a matter of measuring its brightness as seen through filters of different colours, usually blue (B) and 'visual' (V) – yellow-green – filters. Because cool objects glow red, and progressively hotter objects shine orange, yellow, white and blue-white, the method gives the star's temperature quite directly, in terms of a number known as the colour index.

As so often in science, however, the route to a star temperature scale was much more indirect. In late Victorian times, William Huggins in England and Angelo Secchi in Italy produced the first detailed spectra of stars. They found that the dark absorption lines varied markedly from one star to another, and Secchi divided all stars into one of four classes.

In the 1880s, E. C. Pickering at the Harvard College Observatory began to photograph the spectra of thousands of stars. He classified them according to Secchi's system, but with many more subdivisions designated by letters of the alphabet. When the widow of another American astronomer, Henry Draper, gave a bequest to Harvard in memory of her husband, Pickering was able to employ a team of assistants who eventually completed the monumental classification of 225 300 stars in the Henry Draper (HD) catalogue.

The prime mover behind this catalogue was Annie Jump Cannon, who rationalized the order of star types and threw out many of the letters altogether. This gave her the sequence O, B, A, F, G, K, M, R, N, S – which would be impossible to memorize if it were not for the mnemonic 'Oh be a fine girl (guy), kiss me right now – smack!' This mnemonic has now become sadly truncated at 'kiss me', because astronomers currently class the R, N and S stars as subgroups of the M stars. Each class is further divided into ten, so that the Pole Star, for example, is type F8.

The different types of star display lines of different chemical elements, and the Harvard astronomers thought that they were made of different materials. But in the 1920s an Indian astronomer, Meghnad Saha, realized that the lines came from basically the same mixture of gas but raised to different temperatures in different stars. The 'stuff' of ordinary stars consists mainly of hydrogen, with some helium and a little of the other elements. At a temperature of 30 000 degrees Celsius the lines of helium are strong, and we have an O star; at 10 000 degrees Celsius, hydrogen lines are prominent, leading to a classification of A; in a G-type star like the Sun, at 5500 degrees Celsius, the comparatively rare atoms of metals produce the darkest absorption lines.

Another of the Harvard astronomers, Antonia Maury, noticed that even stars of the same spectral type had lines of different widths. The Danish astronomer Ejnar Hertzsprung found that the red stars (types K and M) with narrow lines generally lay further from us than the broad-line stars, and so must be intrinsically brighter and larger. He described the two types as 'giants' and 'dwarfs'. The American astronomer Henry Norris Russell later reached the same conclusion independently. In 1914 the two astronomers published their work jointly in a famous paper in *Nature*, presenting their results in a type of graph that is now fundamental in astronomy, and known as a Hertzsprung–Russell, or H–R, diagram.

An H–R diagram effectively plots the brightness of a star against its temperature. (The coordinates are actually absolute magnitude versus spectral class or colour index.) Most of the stars in an H–R diagram lie along a diagonal, the 'main sequence', that goes from cool dim stars to hot bright stars. Astronomers now know that main-sequence stars are all shining because they are converting hydrogen to helium in their cores. The main sequence is basically an order of mass, with the heavyweight stars at the top of the diagonal band, shining hot and brightly, and the dim lightweight stars at the lower end.

The theory of star evolution tells us that as stars grow older they swell and become cooler, to become red giants – like Betelgeuse or Antares – and move to a region above the main sequence. The most massive stars, at the top of the main sequence, are the first to make the move to the red giant region. Eventually, most stars turn into white dwarfs, which are hot but dim and lie below the main sequence.

The H–R diagram of a cluster of stars or of another galaxy is the astronomer's main tool in unlocking its properties. It allows us to determine some of its most basic properties, such as its distance. From the properties of the stars still left on the main sequence, astronomers can also work out the age of a family of stars that was born together.

If a group of stars contains brilliant massive – but short-lived – O and B stars, then it must have been born in the past 10 million years or so. Such 'OB associations' mark out the stellar nurseries in the spiral arms of our Galaxy and others. In the globular clusters of our Galaxy's halo, on the other hand, all the stars have left the main sequence apart from the longest-lived red dwarfs – so these must be the oldest star groups in our Galaxy. A more detailed analysis of these starry fossils is allowing astronomers to work out just when – and how – our Galaxy originally formed.

Following Baade's advice to seek out the 'candles and frosting' in our Galaxy, the American astronomer William Morgan began the mapping of the Milky Way's spiral structure in 1950. By plotting the positions of brilliant young O and B stars, he found that they lay in two parallel lines, one marking the spiral arm that our Sun lies in and the other the next arm out, the Perseus Arm. Morgan and his students also searched out new nebulae, measuring their distances by investigating the stars that lit them up. The nebulae confirmed the existence of the Local and Perseus Arms, and revealed a new arm – the Sagittarius Arm – closer to the galactic centre.

Other optical astronomers – notably Roberta Humphreys in the United States and Yvon and Yvonne Georgelin in France – have followed up this pioneering work, and plotted associations of OB stars and glowing nebulae over a much wider region. But even the most exacting search for 'optical tracers' of spiral structure takes us out to a distance of only 10 000 light years – less than half the way from the Sun to the galactic

Annie Jump Cannon (1863–1941) pioneered the system of classifying stars according to their spectra. She examined the spectra of nearly 500 000 stars, and between 1918 and 1924, her work was published as the nine-volume Henry Draper Catalogue. Her OBAFGKM series of spectral types, running from hot to cool stars, still forms the basis upon which stars are classified today.

centre. The culprit is the dust in space, which hides the more distant objects. And the dust is thickest in the spiral arms, just where we want to ferret out the stars and nebulae.

To investigate the Galaxy on the widest scale, we need to use the dust-penetrating properties of radio waves. The original search for the 21-centimetre emission from hydrogen was prompted, in fact, not so much to investigate the structure of the interstellar medium as to determine how gas was spread through the Galaxy. For Jan Oort, this would give basic information on the rotation of our Galaxy, because the motion of the gas would show up as a Doppler shift in the wavelength emitted. A plot of the speed against the distance from the galactic centre – the Galaxy's rotation curve – would also provide a means of measuring the distance to farflung gas clouds and nebulae.

In the 1950s, Oort teamed up with Australian astronomers to produce the first overall map of the spiral arms of the Milky Way. In each direction that the radio telescopes in Holland or Australia looked, the astronomers plotted the intensity of the radiation against the speed as shown by the Doppler effect. They assumed each 'peak' in the graph was a hydrogen cloud, and that the Doppler Effect showed its distance. It was then a simple matter to draw a map of all the hydrogen clouds – and, indeed, the map showed a pattern of spiral arms where the hydrogen was most concentrated.

When published in the 1950s, this map of our Galaxy first gave a real feeling that we live in a spiral galaxy. But its overall structure, with many spurs and several apparently double arms did not look quite like other galaxies. And, in the 1970s, it became clear that hydrogen maps do not give an accurate picture of the Galaxy's spiral arms.

The reasons are twofold. First, the hydrogen is not very strongly concentrated towards the spiral arms. Along any line of sight, we look through an immense distance of gas lying

Famous for his collaboration with Henry Norris Russell on the Hertz-sprung–Russell Diagram, the Danish astronomer Ejnar Hertzsprung also discovered that the Pole Star was a Cepheid variable. In 1913, he estimated the distance to the Small Magellanic Cloud by comparing the brightness of its Cepheid variables with nearby Cepheids within our Galaxy. Unfortunately, he published his findings in obscure journals, and so his work – which was ahead of its time – was not fully appreciated.

In 1913, the American astronomer Henry Norris Russell (1877–1957), Harvard's Antonia Maury, and Ejnar Hertzsprung all came independently to the conclusion that the spectra of stars revealed not only their temperature, but also their sizes. The following year, Russell and Hertzsprung published their findings in a classic paper in the journal *Nature*. The graph they included – now known as the Hertzsprung–Russell diagram – has become a fundamental tool for understanding how stars work, and how they change with time.

contain some gas that has been decelerated and some that has not, a simple analysis of the hydrogen data misleadingly shows these arms as being double.

Although the pendulum has now swung against using hydrogen as a tracer of the spiral arms, we should not discount the 21-centimetre data altogether. They still provide a useful indicator of spiral structure in regions with distinct clouds of hydrogen and rather weak effects from the arms – in particular, in the outer parts of the Galaxy. The hydrogen maps show, for example, an Outer Arm that lies beyond the Perseus Arm.

For the inner regions, however, we must turn to other spiral-

Jan Oort – Dutch pioneer of many areas of astronomy, from the distribution of comets to the nature of active galactic nuclei – was the person who initiated the study of the structure of our Galaxy through radio astronomy, mapping the distribution of cold hydrogen gas over the whole sky. The resulting maps showed a convincing spiral pattern, but later work showed that the picture was not as accurate as had originally been thought.

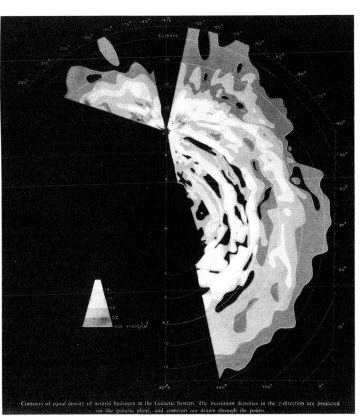

Jan Oort's first radio investigations of the structure of the Galaxy were limited to the view from the northern hemisphere. This, his first map of the Galaxy in the light of the 21-centimetre hydrogen line, shows the distribution of cool hydrogen gas. The densest regions (where there are more than 1.6 atoms per cubic centimetre!) are coded white.

between the arms, and this gas may form a misleading peak in the graph if any slight wind in the interstellar medium disturbs the expected relation between speed and distance. In addition, the arms themselves produce gas motions that can distort the way that we plot them on these maps. According to the density-wave theory, the interstellar gas changes its speed abruptly as it enters the spiral arm – by around 10 kilometres per second. If we naively place such a cloud at the position we associate with the measured velocity, it will be at the wrong distance from the Sun. Because many portions of the arm

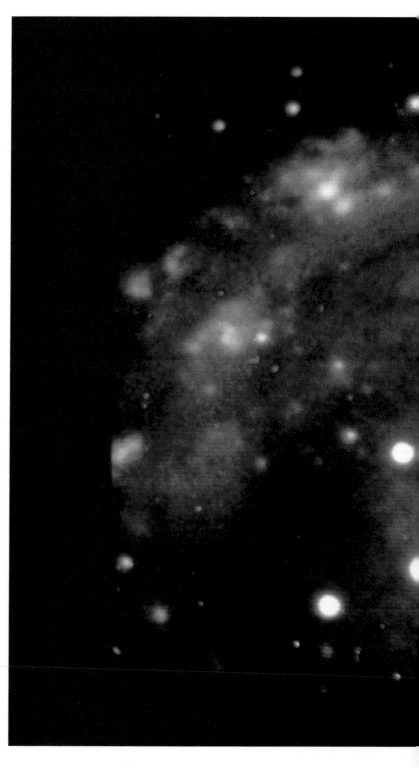

The beautiful spiral galaxy NGC 2997 lies in the southern constellation of Antlia, and is about 30 million light years away. Glowing nebulae – shining pink with H-alpha light from ionized hydrogen – clearly delineate its spiral arms. By studying the nebulae in our own Galaxy and measuring their distances, we can similarly attempt to trace the Milky Way's spiral arms.

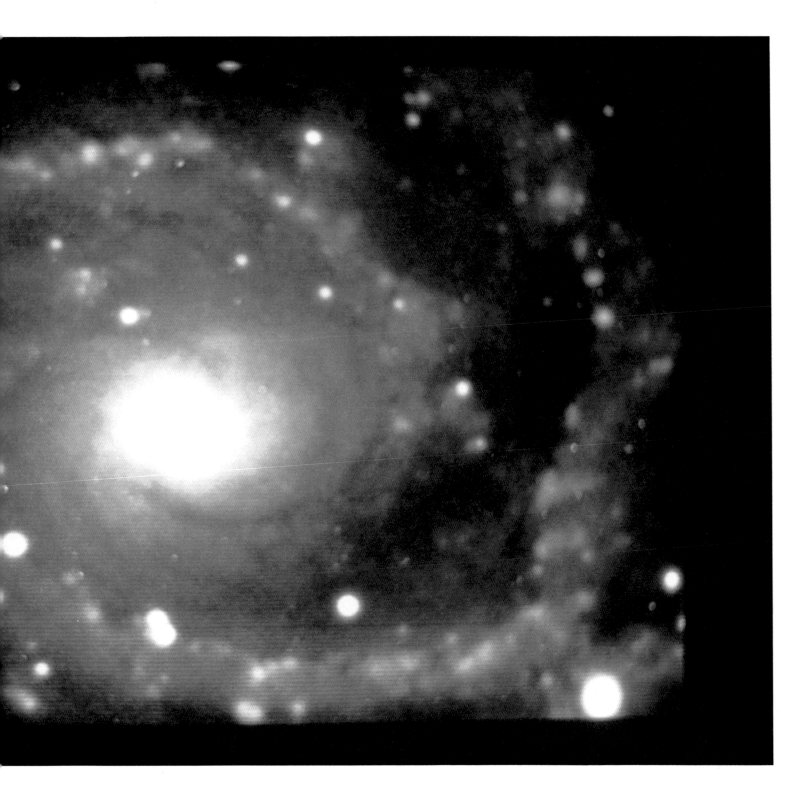

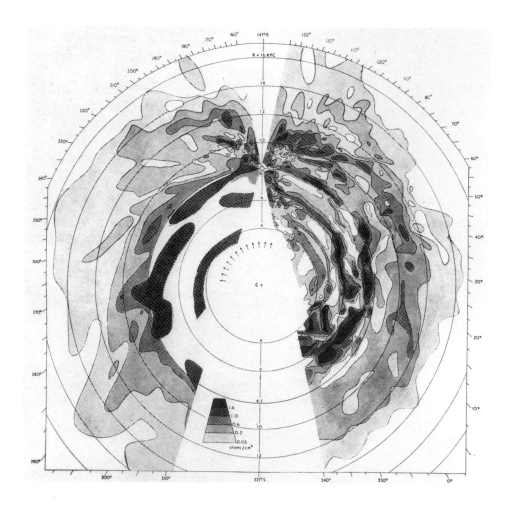

Oort later collaborated with astronomers in Australia to produce this 21-centimetre map of the whole Galaxy. The empty cone is where the hydrogen gas is moving mainly across the line of sight, which does not allow its distance to be estimated. Although this has been superseded by far more accurate maps, it was a remarkable achievement in the 1950s and was an early example of international collaboration in astrophysics.

arm tracers. Nebulae come into their own again here. Radio astronomers can detect nebulae far beyond the limits of optical astronomy, because they emit characteristic spectral lines as electrons recombine with protons (hydrogen nuclei) that have been stripped by the heat. The Doppler effect tells us their distance once again. This time there are no problems with emission along the line of sight or with decelerating shocks at the edge of the arm, because the hot hydrogen occurs only in discrete small clouds within the arms.

In 1976, the Georgelins published what became for many years the standard map of the Galaxy, based on the distances to hot nebulae. They concluded that the nebulae lie along segments of four spiral arms. These do not include the 'Local Arm', which appears as only an insignificant blip between the Perseus and Sagittarius Arms. According to the Georgelins, the Perseus Arm extends round at least one-quarter of the Galaxy, while the Sagittarius Arms wraps itself more than half-way round. Within the Sagittarius Arm, the Georgelins identified two more arms: the Scutum–Crux Arm and the Norma Arm, both covering almost 180 degrees as seen from the centre of the Galaxy.

In the past few years, there has been a chance to improve our knowledge of the geography of the Milky Way – thanks to the recently discovered molecular clouds. There are around 5000 molecular clouds in the Galaxy, and the largest ones occur only in the spiral arms. The strong emisssion from carbon monoxide means that even a small radio telescope can pick up carbon monoxide radiation from a gas cloud right

BOX 3. **Plumbing the depths of the Galaxy**

At first glance, everything in the night sky seems to be at the same distance: pinned to the inside of a giant sphere. The first step in understanding anything about the Universe is to find the distances to the planets, stars, nebulae and galaxies. Only then can we work out how big and bright they are, and begin to untangle the structure of our Galaxy and the Universe on the largest scale.

Beginning near to home, astronomers can find distances in the Solar System by bouncing radar waves off the other planets – and even off the outer atmosphere of the Sun. We have confirmed these measurements by tracking the spacecraft that have visited other worlds. Using the laws of gravity that control the planets' motions, we can derive highly precise distances from the Sun to all the planets – including the Earth.

This provides the first step in the traditional way of finding the distances of the stars: the parallax method. If we look at a nearby star from two different vantage points, it seems to move slightly against the background of more distant stars. The apparent motion due to parallax is larger for the nearest stars, and progressively smaller for more distant objects.

The stars are so far off that astronomers must use the widest pair of vantage points that we have: the position of the Earth on either side of its orbit, separated by almost 300 million kilometres. That means observing a star at intervals six months apart. In practice, it is better to observe over a period of years, partly to average the inevitable errors, and also to separate out the parallax shift of a star from its real motion across space.

Even so, astronomers are looking for something remarkably small. The parallax of the nearest star – Proxima Centauri – is only three-quarters of an arcsecond, equivalent to the size of a small coin at a distance of two kilometres! That's why it took centuries of effort before astronomers measured any star distances at all. The breakthrough came in 1838. Three astronomers cracked the problem almost simultaneously.

First was the German astronomer Friedrich Bessel, who announced that he had measured the parallax of the dim star 61 Cygni. Thomas Henderson, a Scottish astronomer who had been observing in South Africa, then realized that measurements he had made of alpha Centauri would give him the distance to this star. And F.G.W. Struve at the Pulkovo Observatory in Russia measured the distance to the bright star Vega.

These are among the nearest stars, but even so their distances run into millions of millions of kilometres. Astronomers use two larger units of length to avoid the expression of star-distance from becoming truly astronomical. One 'parsec' is the distance at which a star would show a parallax of one arcsecond, and is about 31 million million kilometres. More familiar to most people is the light year – 9.46 million million kilometres – the distance that light travels in one year. Proxima Centauri, the nearest member of the alpha Centauri triple-star family, lies at a distance of 1.29 parsecs or 4.2 light years.

Because the Earth's atmosphere blurs the images of stars, astronomers can measure parallaxes of stars only out to a distance of 300 light years, and that is only for the brighter stars. This radius from the Sun covers only a minute region of our Galaxy (as the maps in Chapter 6 show clearly). In the mid-1990s, astronomers have been able to push out five times further – to cover a hundredfold larger volume of space – using data from the specially designed satellite Hipparcos ('High-Precision Parallax Collecting Satellite'). It has provided a precision catalogue of the distances to 120 000 stars.

As an alternative to the parallax method, astronomers can measure the distance to a cluster of stars in a completely different way. As a cluster moves through space, its stars all travel in parallel lines. The effects of perspective make these lines seem to converge to a definite point in the sky. By plotting the positions of the stars over the years, astronomers can tell what direction the cluster is moving in, and can measure the rate at which it appears to shrink as it moves away from us. A spectroscope can show how fast the stars are moving along the line of sight, towards or away from us. By comparing the cluster's actual speed with the rate at which it seems to shrink, astronomers can work out how far away the cluster must lie.

Astronomers have long used this 'moving cluster method' to measure the distance to the nearest star cluster, the Hyades, which forms the 'head' of Taurus the bull. The Hyades lie some 150 light years away, and the moving cluster method has been more accurate than ground-based parallax measurements. Although Hipparcos can provide an even more accurate distance for the Hyades, based on parallax, the moving cluster method will still be useful for farther clusters.

The next rung up the ladder of distances is to compare distant clusters with those nearby. If we take a star in a far cluster that is identical – according to its spectrum – with a star in, say, the Hyades, then its relative dimness must be caused just by its distance (after making an allowance for absorption by dust in space). In practice, astronomers involve a lot of cluster members at once, by comparing the main sequences in the H–R diagrams of the two clusters: the main sequence of the other cluster should be parallel to the Hyades main sequence, but consistently fainter because of its greater distance.

Within clusters at a known distance, astronomers can find variable stars that are suitable for using as a standard of distance. The type known as RR Lyrae stars, for example, all turn out to have exactly the same intrinsic luminosity (averaged over each star's regularly changing brightness). If we find an RR Lyrae star in another cluster, we can instantly judge the cluster's distance from the apparent brightness of the variable star. This is an ideal way to measure the distances to globular clusters.

Another type of pulsating variable star is the Cepheid: these stars cover a range of luminosities, but the luminosity of any star depends in a precise way on its period. Astronomers have carefully calibrated this period–luminosity relationship for Cepheids at a known distance, so they can now find the distance of any other Cepheid simply by measuring its period and its apparent brightness.

The data amassed by astronomers in these ways then allow them to achieve the amazing feat of finding the distance to any ordinary star, in a cluster or on its own in space. Over the past few decades, astronomers have studied the spectra of thousands of stars with known distances. For a star of any particular spectral type, there is a correlation between the width of the spectral lines and the star's luminosity. The underlying reason is that a brighter star is larger than a dimmer star with the same temperature: as a result, the gravity pulling on its outer layers is lower, giving a lower pressure, and gas at a lower pressure produces narrower spectral lines.

Leaving aside the detailed physics, this gives us an ideal method for measuring star distances (a method known, rather confusingly, as spectroscopic parallax). Researchers have tabulated the luminosity for stars with any combination of spectral type and line-width. If we are investigating a star for the first time, we simply check its spectrum and look up the luminosity it should have: by comparing this with the apparent brightness of the star in the sky, we can readily work out the distance.

Star distances provide the backbone to plumbing the depths of our Galaxy. If we come across a glowing nebula or a dark cloud, for example, the most accurate way to judge its distance is to find a star that is immersed in the gas and dust, and determine its distance by spectroscopic parallax. In this way, astronomers have mapped out our local region in space, and the two nearest spiral arms of our Galaxy.

But the dust in the Milky Way prevents us from seeing right across our Galaxy: indeed, we can only map the stars in a region a few thousand light years across. Astronomers can detect more distant objects with infrared telescopes and radio telescopes that can see through the dust. They reveal thousands of more distant hot nebulae and dense, star-forming clouds – with no easy way to work out their distances.

With these remote objects, astronomers must turn to using the rotation of our Galaxy. Like the Sun and the stars, these gas clouds are moving around the galactic centre at high speed. But they do not stay the same distance from us, as they would if our Galaxy were rotating like a solid body such as a car wheel. The Galaxy is held together by gravity, and its influence becomes weaker as you move outwards, leading to slower speeds in the outer regions – rather like the Solar System, where Pluto moves much more slowly than Mercury.

As a result, the distant gas clouds are, in general, moving along our line of sight. Radio astronomers can tune into a spectral line in the cloud, coming from either hydrogen or a molecule like carbon monoxide, and the motion of the gas shows up as a Doppler shift in the wavelength of the radiation. There are four special directions – spaced by 90 degrees in the sky – in which we observe no Doppler Effect, because the gas clouds always keep at the same distance from the Sun: directly towards the centre of the Galaxy and directly away; and along the Sun's orbit, to the front and to the rear.

In all other directions, we observe matter that is moving either towards or away from us – and its speed depends on its distance. This method (known rather illogically as kinematic parallax) does, however, suffer a problem if we are looking at objects within the Sun's orbit around the Galaxy. Any particular speed will correspond to two different distances along the line of sight. With increasing distance from the Sun, the speeds (relative to us) increase steadily until we reach the point along the line of sight that is closest to the Galaxy's centre. The speed then decreases again, until it reaches zero once more for an object that is following the same orbit as the Sun but further round the Galaxy. In practice, astronomers have to look at all such objects individually to search out clues to whether they actually lie at the 'kinematic near distance' or the 'kinematic far distance'.

With these tools, astronomers have, in recent years, been able to apply precise cartography to the stars in our local neighbourhood and to draw up a general plan of even the most distant reaches of the Galaxy, almost 200 000 light years away. It is sobering to realize the progress made in just over 150 years, since the first hard-won parallax measurements on stars only a few light years from us.

across the Galaxy. Indeed, the problem of discovering new molecular clouds was originally that the existing large radio telescopes focused on too small an area of sky: it would have taken decades to scan the sky to build up a complete map of carbon monoxide emission.

In 1975, Patrick Thaddeus at Columbia University built an exceptionally small radio telescope, only 1.2 metres across, to search out the molecular clouds of the Milky Way. It was perched on top of a university building in New York City. Although pollution would wipe out astronomy in the optical and at most radio wavelengths, at 2.6 millimetres New York is, according to Thaddeus, 'as quiet as the day Henry Hudson first sailed up the river'. Its survey of the northern sky was matched with a similar telescope in Chile to sweep the southern heavens.

Thaddeus found that the molecular clouds in our Galaxy are largely concentrated in a ring of gas that lies 15 000 light years from the galactic centre. From this ring spring the beginnings of the Scutum–Crux and the Norma Arms. Here, the individual clouds are so crowded that it is difficult to work out the overall structure.

The V-shaped cluster of the Hyades – the 'head' of Taurus the Bull – is the nearest open cluster to the Sun, and one of the cornerstones of the distance scale. By studying the way in which the stars in the cluster move (the 'Moving Cluster' Method), astronomers can measure the distance to the Hyades, and thereby determine the luminosity and temperature of the stars in the cluster. Then, by comparing stars in more distant, but similar, clusters with the stars in the Hyades, astronomers can determine how far away the more remote clusters are. At a distance of 150 light years, the Hyades themselves are at the limit of the range of parallax. (Aldebaran – the lower 'point' of the V – is not a member of the cluster, and lies a lot closer at 68 light years.)

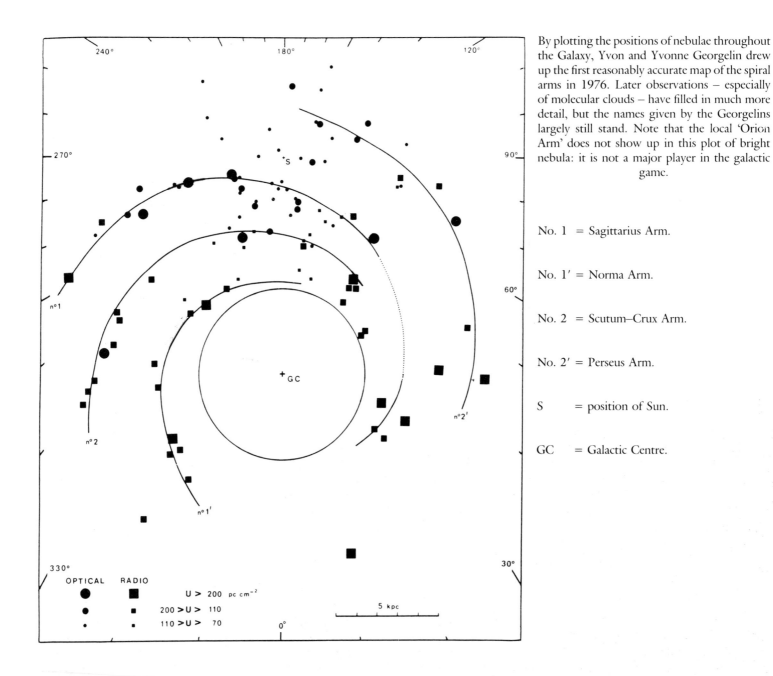

By plotting the positions of nebulae throughout the Galaxy, Yvon and Yvonne Georgelin drew up the first reasonably accurate map of the spiral arms in 1976. Later observations – especially of molecular clouds – have filled in much more detail, but the names given by the Georgelins largely still stand. Note that the local 'Orion Arm' does not show up in this plot of bright nebula: it is not a major player in the galactic game.

No. 1 = Sagittarius Arm.

No. 1' = Norma Arm.

No. 2 = Scutum–Crux Arm.

No. 2' = Perseus Arm.

S   = position of Sun.

GC   = Galactic Centre.

Outside the dense molecular ring, however, Thaddeus and his colleagues have been able to separate the individual clouds along any line of sight by their different velocities. And these speeds have indicated the distance of the molecular clouds from the Sun. As a result, the team has been able to build up a view of the spiral arms that is unprecedented in its detail and accuracy.

The Sagittarius Arm turns out to be a major arm of the Galaxy. So does the Outer Arm – known from hydrogen observations – which winds inwards for at least an equal length.

The world's smallest radio telescope? Patrick Thaddeus built this 1.2-metre dish on the roof of a Columbia University building in New York City to study the molecular clouds of the Milky Way. Although terrestrial interference in a city the size of New York would normally be expected to drown any signals from space, at 2.6 millimetres – the wavelength Thaddeus chose – there is very little background noise. In 1986, the dish moved with Thaddeus' team to the Harvard–Smithsonian Centre for Astrophysics in Cambridge, Massachusetts.

Other 'arms' turn out to be merely shorter segments. These include the Perseus Arm, as well as our own Local – or Orion – Arm. On the local scale, the carbon monoxide observations reveal that the dark dust clouds running from Cygnus through to the Coal Sack constitute a band of dust similar to the dust we see strung along the inner edge of the arms of other galaxies.

These observations of molecular clouds have brought us a major step closer to tracing the structure of our Galaxy. We still have to be somewhat cautious at interpreting exactly where the molecular clouds actually lie, but the overall view is now becoming clear.

Many of the early interpretations of the spiral arms were flawed because they tried to join up all known arm segments into a few major spiral arms. When two arms obviously did not suffice, astronomers opted for four. They sometimes even tried to force the observed spiral-arm tracers to fit precise mathematical spiral patterns. As early as 1983, however, the American radio astronomer Frank Kerr had commented presciently: 'one guideline has been used by many people, namely that they look for a very *regular* spiral pattern. However, this guideline must be wrong, as no other galaxies are so regular'.

With the huge number of molecular clouds now known, we do not have to resort to mathematical fits. A map of the clouds themselves shows the overall shape of the Galaxy. Our basic diagram (Map A, p. 54) simply plots the positions of all the reliable spiral-arm tracers, with data from the IRAS satellite on the distribution of stars near the centre of the Galaxy. Using the best pattern-recognition device known to science – the human eye and brain – it is easy to pick out some major segments of arm, and to realize that there are regions where a simple spiral pattern does not fit at all.

We have gone also one step further, by blacking in the regions between the obvious spiral arm segments. Although this procedure is somewhat subjective, its main effect is simply to highlight the pattern that already exists. From this skeleton, we have produced a simple 'realistic' view of how our Galaxy would actually look from the outside (Map B, p. 55). Interestingly enough, it also produces a Milky Way Galaxy that has a similar feel to photographs of other spiral galaxies.

Our map shows that the Galaxy has two major spiral arms, presumably the result of a density wave that sweeps around the entire Galaxy. One is the Norma Arm, which in our recon-

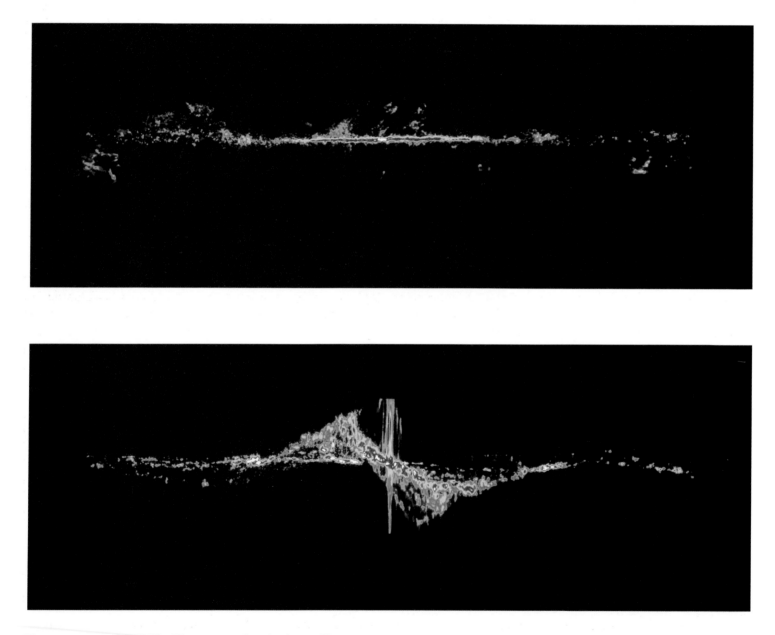

These two maps of the Milky Way at a wavelength of 2.6 millimetres were made by twin 1.2-metre radio telescopes – one in New York, the other in Chile. The top map shows the distribution of carbon monoxide molecules in our Galaxy, coded for intensity, with the galactic centre in the middle of the map. The bottom image reveals both the distribution and velocity of the molecular clouds (the vertical scale measures velocity: positive if towards us, negative if away). Most of the velocity spread is caused by different parts of the Galaxy travelling at different speeds (differential rotation), but the very high speeds close to the galactic centre may be caused by outflows. The ring of molecular clouds close to the galactic centre shows up very clearly.

BOX 4. **The Sun's location and motion**

In 1918, the American astronomer Harlow Shapley found that the Sun does not lie in the centre of the Milky Way Galaxy. Since then, astronomers have debated our precise location – and how the Sun is moving.

The first crucial measurement is to find the distance from the Sun to the galactic centre. For many decades, astronomers had only a hazy idea, and they adopted a round figure of 10 kiloparsecs (about 33 000 light years). Recent measurements (see Chapter 8), however, put the Sun at only 25 000 light years from the centre of the Galaxy.

This measurement is bound up with the question of the overall size of the Galaxy. Astronomers measure the distance to the far reaches of our Galaxy in terms of the Sun's distance from the centre. Early researchers put the Sun about two-thirds of the way out, which made the Galaxy as a whole 100 000 light years across. Even though our distance from the galactic centre has now shrunk, astronomers have found further-flung stars and gas clouds – some twice the Sun's distance from the Galaxy's heart. As a result, we can still adopt a round figure of 100 000 light years for the Galaxy's diameter.

Once we know the distance from the Sun to the galactic centre, we can use measurements of gas in the rotating Galaxy to deduce the actual rotation rate at different distances from the galactic centre. At the Sun's distance, this works out to be 200 kilometres per second – a figure that trips glibly off the tongue, but in fact represents an unimaginable three-quarters of a million kilometres per hour. The Galaxy is so vast that even travelling at this headlong rate, the Sun will take 240 million years to complete one orbit around the Galaxy – one 'cosmic year'.

All the stars around the Sun (apart from a few halo stars currently passing through the disc) are travelling at much the same rate. Despite their speedy motion around the Galaxy, the stars do not move very fast relative to one another – just as cars in the fast lane of a motorway keep much the same distance apart even though they are speeding along the road.

Astronomers call the average motion of all the stars around us 'the local standard of rest': somewhat paradoxically, as it is actually moving at 200 kilometres per second in a circular path around the Galaxy. By definition, the local standard of rest has no component of motion in the other two directions: towards or away from the galactic centre, and up or down in the galactic plane.

Like cars on a motorway, all the stars in the disc have motions that differ slightly from the local standard of rest. The Sun is no exception. We can find out how our star is moving by measuring the apparent motions of many nearby stars. If the Sun were stationary, relative to the local standard of rest, then these motions would cancel out to zero. In fact, the average motion of our neighbours is not zero: the nearby stars seem to stream by as we move past them.

As early as 1783, William Herschel deduced that the Sun was on the move. By analysing the apparent motions of 14 bright stars, he was able to put the direction of the Sun's motion – the 'solar apex' – in the direction of the star lambda Herculis. More recent results have confirmed that the Sun is indeed moving towards Hercules, at a speed of 20 kilometres per second.

Compared to the local standard of rest, the Sun is moving faster around the Galaxy by 14 kilometres per second; it is also travelling in towards the galactic centre at 10 kilometres per second. The American astronomer Frank Bash has interpreted these figures to mean that the Sun is in a roughly elliptical orbit where its far point (apogalacticon) is seven per cent further away than its current location. The Sun is now travelling inwards and has almost reached its nearest point to the galactic centre, the perigalacticon, which lies at 99.5 per cent of the Sun's current distance from the galactic centre. The Sun will reach perigalacticon in only 15 million years' time – less than a month in terms of the cosmic year.

The Sun is also moving upwards, out of the plane of the Milky Way, at a speed of seven kilometres per second. At the moment, the Sun is 50 light years above the midplane of the Galaxy, and its motion is steadily carrying it further away. But the gravitational pull of the stars in the plane is slowing down the Sun's escape. Bash calculates that in 14 million years from now – fortuitously, about the same time as reaching perigalacticon – the Sun will reach its maximum height above the galactic disc.

From this perch, 250 light years above the Galaxy's midplane, the Sun will be pulled back towards the galactic disc.

Passing through, it will travel to a point 250 light years below the disc, and then oscillate upwards again to reach its present position again 66 million years from now.

In its perpetual bobbing up and down, the Sun crosses the plane of the Galaxy every 33 million years. Some scientists have suggested that the plane-crossings coincide with periods of mass extinction on the Earth – such as the death of the dinosaurs 66 million years ago – but the evidence is very weak. We crossed the plane two million years ago, for example, and there's no sign of major damage to our planet then.

The motion of the Sun does, however, affect our view of the Galaxy. As we are currently in the thick of the Galaxy's disc, our view of the distant regions is largely blocked by dust. Riding the solar roller-coaster to its maximum height, our descendants of 10 to 20 million years' time can expect a grandstand view of our star-city.

struction passes behind the galactic centre and reemerges as the Outer Arm – in all, making more than one complete turn around the Galaxy. The Sagittarius Arm near the Sun is the second of these major arms. It springs from one end of the stubby bar in the Galaxy's centre, on the opposite side from the Sun. Some 180 degrees on, it passes just within the Sun's position. Here lie the giant Lagoon and Omega Nebulae, while even more massive gas clouds – the Carina Nebula and NGC 3603 – mark its further extensions. This arm fades out after making one complete turn around the Galaxy.

Other parts of the Galaxy are broken up into a plethora of spiral segments and smaller fragments, forming flocculent regions between the main arms and beyond them. In its texture, then, as in everything else, our Galaxy turns out to be a middling creature.

To get a good idea of the outer regions of the Milky Way, we can turn to photographs of M83 in Hydra. This galaxy is an open Sc type and is much more active than the Milky Way, with spiral arms festooned with glowing nebulae: since 1923,

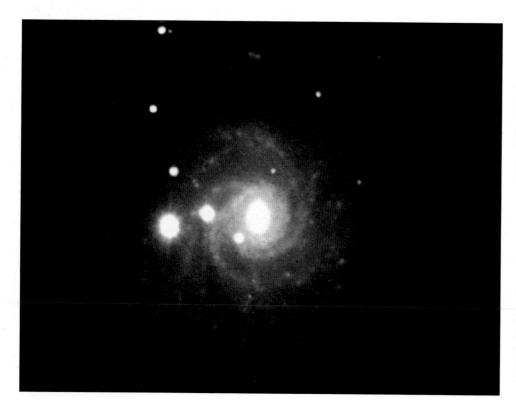

NGC 3344 is a galaxy that may bear a very strong resemblence to our own. It lies about 35 million light years away in Leo Minor, and its morphological classification comes very close to that of the Milky Way on de Vaucouleurs's scheme. Additionally, it has a prominent ring of nebulae surrounding its nuclear bulge – corresponding to the molecular ring in our own Galaxy – and, like the Milky Way, its spiral arms are long but patchy. (The very bright stars in the image are foreground stars in our Galaxy.)

FACING PAGE The outer parts of our Galaxy may resemble the glorious spiral arms of M83. Decked out with glowing nebulae, M83 is vigorously producing new stars – some of which have exploded as supernovae. Astronomers have logged five exploding stars in this galaxy since 1923. Although the Milky Way is not such an active galaxy, its outer arms are certain to bear a family resemblence to those of M83, with its flocculent clumps of gas and dust and sparkling clusters of stars.

astronomers have seen a record five supernovae explode in M83. Using the Anglo-Australian Telescope in New South Wales, the leading astrophotographer David Malin has taken colour pictures that show clearly the details in the patchy spiral arms of M83: the clouds of pink gas, the scattered blue OB associations and the streamers of dark dust that must surround us in the outer arms of the Milky Way.

No nearby galaxy is an exact match to the image of the Milky Way that we have built up here. One reasonable match lies in Leo Minor, about 35 million light years away. NGC 3344 has been described by different authors as SBbc(rs) and SAB(r)bc – as close to de Vaucouleurs's classification of the Milky Way as we could hope to come. Surrounding this galaxy's centre is a ring of nebulae that matches the molecular ring in the Milky Way; winding outwards are two long but patchy arms, with shorter arm segments between them.

A few years ago, the leading Galaxy experts Bart and Priscilla Bok concluded that 'the game of connecting recognized radio and optical spiral features into an overall pattern for our Galaxy will undoubtedly continue for some time'. That statement still holds true for the smaller features of the Milky Way. But after the recent mapping of giant molecular clouds, astronomers are confident that we are beginning to understand the basic structure of the Galaxy.

The image of the Milky Way that we present in this book is the best that the astronomy of the 1990s can achieve – and its bare bones are certainly close to the true picture of the Galaxy. This is how an inhabitant of Andromeda would see our home in the Universe.

CHAPTER 4

The Perseus Arm

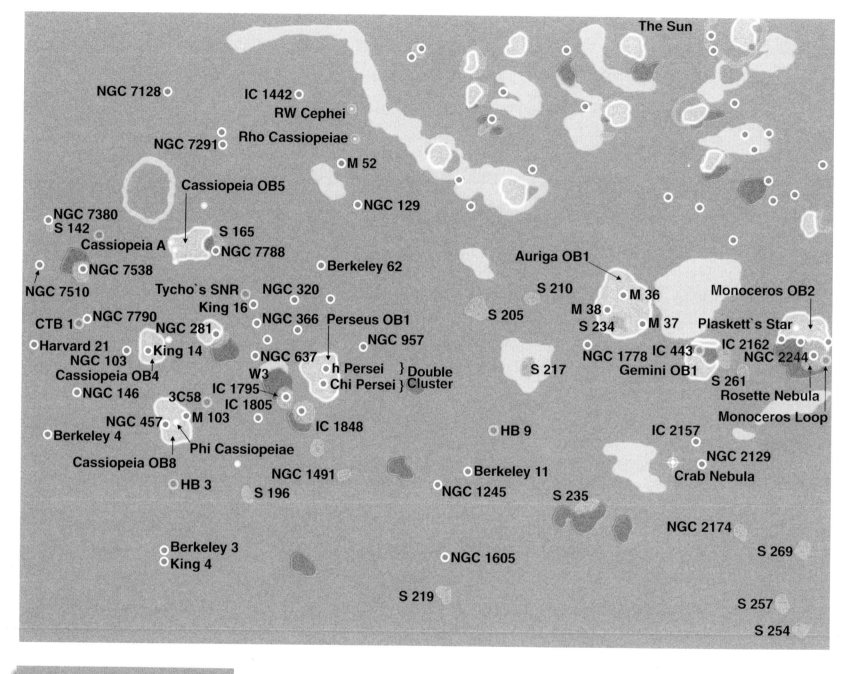

The Sun

NGC 7128   IC 1442
                  RW Cephei
NGC 7291   Rho Cassiopeiae
                       M 52
          Cassiopeia OB5
                                  NGC 129
NGC 7380
S 142   S 165
Cassiopeia A   NGC 7788
NGC 7538   Auriga OB1
                             Berkeley 62
NGC 7510   S 210
        Tycho`s SNR   NGC 320   M 36   Monoceros OB2
CTB 1   NGC 7790   King 16   S 205   M 38
                        NGC 366   Perseus OB1   S 234   M 37   Plaskett`s Star
        NGC 281   NGC 957   IC 2162
Harvard 21   King 14   S 217   NGC 1778 IC 443   NGC 2244
NGC 103   NGC 637   Gemini OB1
Cassiopeia OB4   W3   h Persei } Double   S 261
NGC 146   3C58   IC 1795   Chi Persei } Cluster   Rosette Nebula
NGC 457   M 103   IC 1805   Monoceros Loop
Berkeley 4   IC 1848   HB 9   IC 2157
        Phi Cassiopeiae   NGC 2129
Cassiopeia OB8   Crab Nebula
          HB 3   NGC 1491
                    S 196   Berkeley 11
                              NGC 1245   S 235
                                                      NGC 2174
Berkeley 3   S 269
King 4   NGC 1605
          S 219
                                                              S 257

                                                              S 254

500 Light Years

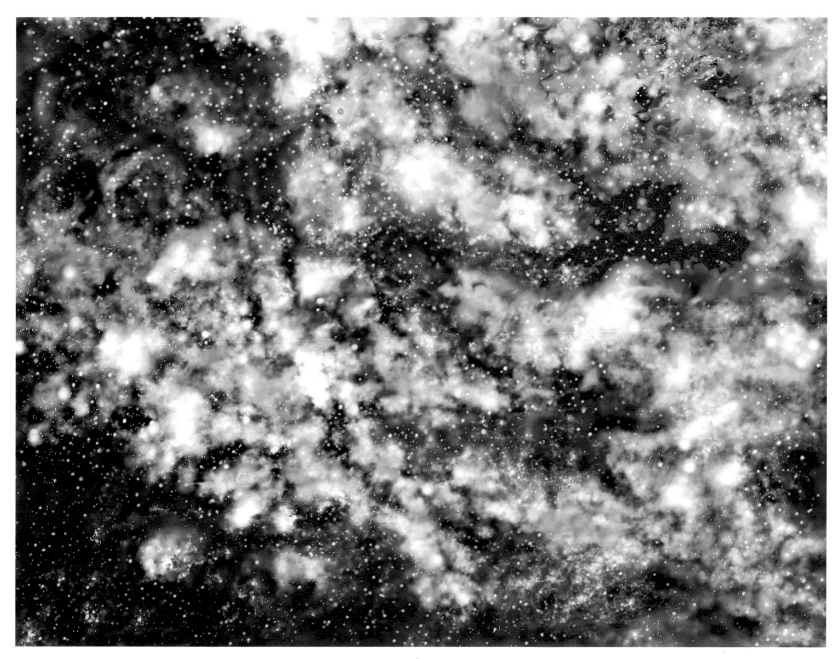

KEY TO MAP A OVERLEAF

KEY TO MAP A

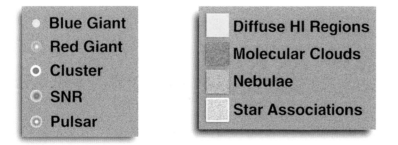

- ● Blue Giant
- ◉ Red Giant
- ○ Cluster
- ◎ SNR
- ◉ Pulsar

- □ Diffuse HI Regions
- ■ Molecular Clouds
- □ Nebulae
- □ Star Associations

MAP A Young clusters and star associations delineate the Perseus Arm. It also contains young supernova remnants, such as Cassiopeia A and the Crab Nebula, that are among the most powerful in the Galaxy.

MAP B The Perseus Arm is fairly broad and ragged, with a pronounced gap near the Crab Nebula. In places (top right) it almost merges with the Local Arm.

In the autumn of 1951, a leading American astronomer happened to look at the sky – and experienced a revelation into the nature of our Galaxy. It was an unexpected insight. Since the beginning of this century, astronomers have turned away from looking directly at the sky. Instead, they have pored over photographs (and more recently, electronic results) taken by telescopes that operate more or less automatically. Only a few professional astronomers can actually recognize one star in the sky from another.

William Morgan was one of the few. In 1951, he was working at the Yerkes Observatory, near Chicago, trying to work out whether our Galaxy has a spiral structure, like that of the Andromeda Galaxy. His project involved measuring the distances to hundreds of stars, of spectral types O and B – the kind that we see concentrated into the spiral arms of other galaxies. Most of these stars were quite close to the Sun, and would clearly lie in the same spiral arm as our star. Where was

he to find the much more distant stars that would lie in the *other* spiral arms, thousands of light years away?

'I was walking between the observatory and home, which is only 100 yards away', Morgan later recalled. 'I was looking up in the northern sky, just looking up in the region of the Double Cluster, when it suddenly occurred to me that the Double Cluster in Perseus and then a number of stars in Cassiopeia and even Cepheus, that along there I was getting distance moduli of between 11 and 12'. In more conventional units, these 'distance moduli' meant that all these stars lay between 5000 and 8000 light years away.

Morgan couldn't wait to get back to the observatory, and plot where these stars were in space. The O and B stars in Perseus and Cassiopeia fell along a narrow 'lane'. Was this a real arm? That rather depended on whether the lane was distinct from the regions where the other stars lay. So Morgan then plotted the other O and B stars in his list. These all fell

In 1951, the American astronomer William Morgan – pictured here – became the first person to identify one of the spiral arms of our Galaxy. Morgan discovered the Perseus Arm by realizing that the brighter stars in Perseus and Cassiopeia all lay at roughly the same distance. Here he uses a hand-lens to examine photographic negatives mounted on a light-box.

much closer to home: 'through the Sun there was this narrow lane parallel to the other one'. Between the two, there was a gap devoid of stars. Morgan's two 'lanes' were undoubtedly portions of two spiral arms in the Milky Way Galaxy. The 'lane' through the Sun is now known as the Local Arm, or Orion Arm; the further arm – the first identified spiral arm of our Galaxy – is called the Perseus Arm.

Morgan extended his quest for spiral arms by searching for bright nebulae, of the kind that we see strung along the arms of other spiral galaxies. These again fell into the well-defined regions of the Local Arm and the Perseus Arm. Morgan presented his results at a meeting in Cleveland, Ohio, at the end of 1951. One participant, Otto Struve, later recalled the response. 'Astronomers are usually of a quiet and introspective dispo-

---

BOX 1. **Waves of gas**

The Perseus Arm is the best place in our Galaxy for us to study how gas moves as it circles the galaxy and plunges into a spiral arm. We detect the hydrogen gas with radio telescopes tuned to a wavelength of 21 centimetres. The wavelength emitted by any particular cloud of hydrogen is affected by its speed, and we can use this to calculate the distance to the cloud (see Chapter 3).

If we could view our Galaxy from the northern regions of intergalactic space we would see it rotating clockwise. This means that the Sun is catching up with the part of the Perseus Arm in the direction of Cassiopeia and Perseus (the region we are considering in this box). Relative to us, the gas in the Perseus Arm seems to be approaching, and has a 'blue-shift'.

When radio astronomers first measured the emission from the gas here, they did indeed find blue-shifted hydrogen gas. But they were surprised to discover that the gas in the Perseus Arm did not all move at the same speed. In some directions, there were gas clouds approaching us rather faster. The early radio astronomers naturally assumed that the higher-speed gas lay further from us, so they concluded that the Perseus Arm was actually double.

This result, unfortunately, conflicted with what the optical astronomers had to say. They had found that all the young stars in this part of the Perseus Arm lie at much the same distance, forming a single arm about 3000 light years broad. In addition, they found absorption lines from the higher-speed interstellar gas in the spectra of stars that were definitely at the nearer edge of the Arm. So the high-speed gas could not lie beyond the Perseus Arm proper. It was in the Arm, or even slightly closer to us.

So what gave this gas its extra speed towards us? Ingenious explanations were invoked: gas in the arm had been propelled towards us by the combined radiation of all the energetic young stars in the Arm, or by the explosion of a 'super-supernova'.

But in 1972, an American astronomer, William Roberts, came up with an explanation that was at once simple and natural. He was an early proponent of the idea that the spiral structure of a galaxy is a result of a density wave: a pattern that keeps its shape and integrity, as stars and gas continually passed into a spiral arm region and out again. One consequence is that gas piles up in the spiral arms, and new gas moving in will run into the gas already in residence. The impact creates a shock wave, which slows down the incoming gas.

In the case of the Perseus Arm, the gas approaches from the inside – roughly from the direction of the Sun. The shock wave slows this incoming gas abruptly.

To work out what we measure, according to this theory, we must recall that all the stars, including the Sun, are whirling around in our giant carousel of a galaxy. When the shocked gas entering the Perseus Arm slows down, relative to space as a whole, it means that the Sun catches up with it faster than we are chasing the gas in the Perseus Arm. So the shocked gas has a higher speed towards the Sun – exactly what we observe in the strange gas clouds that lie in front of the main stars of the Perseus Arm.

So the apparently odd behaviour of hydrogen gas in the Perseus Arm has turned out to be a natural consequence of our Galaxy's spiral nature. And these measurements are helping astronomers to understand better the nature of spiral galaxies in general.

sition. They are not given to displays of emotion . . . But in Cleveland, Morgan's paper on galactic structure was greeted by an ovation such as I have never before witnessed'. Morgan was surprised, too: 'people started to applaud by clapping their hands, but then they started stamping their feet. It was quite an experience'.

Although our plan of the Galaxy has now expanded far behind these two arms, the Perseus Arm is still a touchstone for understanding the spiral arms of the Milky Way. It is difficult to work out the overall shape and motions in the Local Arm, because we are buried in the midst of it, and it is not so easy to unravel our other neighbouring segment of spiral structure, the Sagittarius Arm, because we see it in front of the great complicated structures towards the centre of the Galaxy. There is little behind the Perseus Arm to confuse us, however, because in this direction we look towards the edge of the Milky Way.

For radio astronomers, too, there are advantages in studying the Perseus Arm. They use the apparent speed of a gas cloud to determine its distance (as described in the previous chapter). Unfortunately, the speed of a gas cloud that lies in a direction towards the centre of the Galaxy can apply to two very different distances – one on the near side, and the other on the far side, of the galactic centre. In the direction of Perseus, however, there is no such ambiguity, and gas in the Perseus Arm has yielded new insights into the way that our Galaxy maintains its spiral shape.

Finally, the Perseus Arm has been unusually important for a purely parochial reason: until recent years most telescopes were based in the Earth's northern hemisphere – regions that give us a grandstand view of Perseus.

The Double Cluster is a good place to begin our exploration of the Perseus Arm. It consists of two distinct clusters of stars, both very bright. They owe their brilliance partly to the fact

The twin star clusters of h and chi Persei are easily visible to the naked eye in northern skies, and are one of the most beautiful sights through binoculars or a small telescope. The clusters, which are 7000 light years away, each contain several thousand stars and are separated by 50 light years. Although they look identical, it now appears that h Persei (right) is slightly larger and older (five million years), whereas the stars in chi Persei are only three million years old. The clusters form the heart of an OB association that extends for 750 light years.

that each consists of many thousands of stars, and also to the fact that they are comparatively young. As a result, they contain many of the furiously burning O and B stars that are blue-white in colour and live for only a few million years. We can therefore see the two clusters easily with the naked eye, even though they lie some 7000 light years away.

On a clear night, you can spot the Double Cluster as two small misty patches of light, about half a degree apart, between Perseus and Cassiopeia. On old star charts, they mark the hand of the hero Perseus that is holding his sword above his head. The Greek astronomer Hipparchus recorded them two thousand years ago as a 'cloudy spot'. When the seventeenth-century astronomer Johann Bayer assigned letters to the stars in every constellation, he gave each of the clusters an assignation appropriate to a single star, 'h Persei' and 'chi Persei'.

Oddly enough, the clusters do not appear in Charles Messier's great catalogue of nebulous objects, even though each cluster is brighter than all except four of the objects that Messier did include. This is rather a pity, because many popular books that single out the 'Messier objects' thereby miss out on a pair of the most brilliant sights in the northern sky.

Each of the clusters is about the same apparent size as the Full Moon. A moderate-sized telescope shows about 300 stars in each – the brightest of a total population of several thousand. The most brilliant stars are blue-white O and B types, but several red giants are also scattered through the clusters.

Although the clusters appear as nearly identical twins, they are not exactly the same age. The latest measurements suggest that h Persei is slightly the older, at five million years, while chi Persei is only three million years old. The two clusters form the heart of an 'OB association' of young stars, that stretches for 750 light years around the two clusters.

The Double Cluster lies about 500 light years 'below' the plane of the Milky Way Galaxy. Almost directly above them, in the plane of the Galaxy and in the constellation Cassiopeia, is another pair of clusters that are much younger. IC 1805 and IC 1848 are only about a million years old, and the brightest stars in each are similar to the most brilliant stars in the Double Cluster. But these clusters contain far fewer stars, more spread out, so they are just below the limit of naked eye visibility. Long-exposure photographs show that these two clusters are surrounded by glowing shells of gas, the material left over from their births.

Photographs of this region also show a strange triangular-shaped nebula next door to the two clusters and their tatters of gas. This object, IC 1795, is the lit-up corner of a large dark cloud. Within the cloud, stars are being born right now, and this region is the largest 'star factory' in the Perseus Arm. Although the dust in the cloud hides most of the light from the stars buried within, their radiation at longer wavelengths escapes freely. Hence the dark cloud is a brilliant object when observed at infrared and radio wavelengths. Astronomers call it 'W3', after its number in a catalogue of radio sources compiled by the Dutch astronomer Gart Westerhout in 1958.

Radio and infrared telescopes show us that the dense interstellar cloud W3 contains several stars more brilliant than 100 000 suns. The intense radiation from these brilliant young stars is pushing back the gas and dust immediately around the stars themselves, so that each star is surrounded by a 'shell' of hot compressed matter. Eventually – perhaps in only a thousand years – the radiation from some of these stars will burn through the dust in W3, and we will see the stars directly. They will be easily visible to the naked eye, so changing the appearance of the familiar constellation Cassiopeia. At present, we see Cassiopeia as a 'W' shape of five bright stars: when the brightest star in W3 becomes visible, it will extend this shape to a six-star zigzag.

The five stars of Cassiopeia that appear bright in our skies are all relatively close to the Sun: less than a thousand light years away. Behind them lie many other regions of star-formation and young star clusters, which continue the line of the Perseus Arm as it winds around the Galaxy.

A signpost to the next cluster of young stars is phi Cassiopeiae, a naked-eye star (magnitude 5) that lies below the characteristic W-shape. This is a brilliant yellow supergiant star, hundreds of times larger than the Sun. Shining with the brightness of 200 000 suns, phi Cassiopeiae is one of the most luminous stars in the Galaxy – which is why we can see it with the unaided eye even though it lies over 9000 light years away.

With binoculars or a telescope, you find that phi Cassiopeiae is the brightest star in a cluster, NGC 457. This group has acted as an 'ink-blot test' for amateur astronomers. Referring to the sight seen through binoculars, an American writer, Alan MacRobert, says: 'viewing the cluster for the first time this way on a wild and windy October night a few years ago, I imagined it to be a wisp of candle flame blown . . . by the

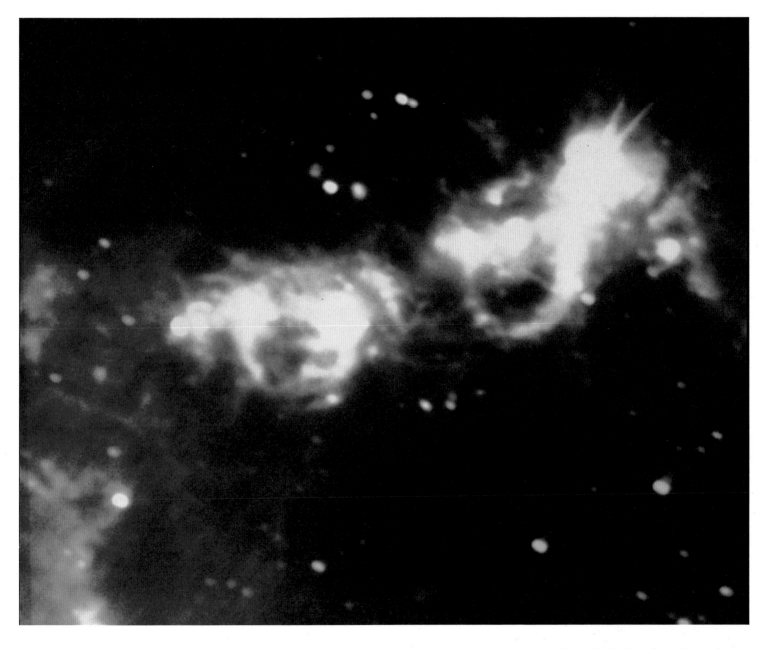

Above the Perseus Double Cluster, in the constellation of Cassiopeia, is another double cluster: IC 1805 (right) and IC 1848 (left). These clusters are much sparser than h and chi Persei, and, at only a million years old, they are younger. Because the clusters are so young, they still contain considerable amounts of their natal dust and gas – which can be seen wreathed around the stars in this infrared image from the IRAS satellite.

The star cluster NGC 457 is 25 million years old and lies just below the 'W' of Cassiopeia. Its most prominent star is the yellow supergiant phi Cassiopeiae, which is visible to the unaided eye, even though it lies 9000 light years away. Phi Cassiopeiae is 200 000 times brighter than the Sun, and will almost certainly explode as a supernova within the next few thousand years.

FACING PAGE On the edge of IC 1805 is a large dark cloud – W3 – which turns out to be the biggest star-forming region in the Perseus Arm. The nebula IC 1795 marks the only – as yet – lit up part of it. Infrared and radio waves penetrate the dust, however, and this infrared view reveals that star-birth is actively in progress. Some of the young stars are brighter than 100 000 suns, and when their light burns through the dust – in perhaps a thousand years – the characteristic W-shape of Cassiopeia will be radically altered.

wind'. Through a telescope, MacRobert sees the centre of NGC 457 as a long hollow arrowhead. To other observers, the cluster looks like an owl, or even the film alien E.T.

NGC 457 is about 25 million years old. The brilliant super-giant phi Cassiopeia has swollen to its current vast size as it nears the end of its life. Before many more thousand years have passed, it will erupt as a supernova: as we'll see below, it will be just the latest in a long line of exploding stars in this region of the Galaxy.

Not far from NGC 457, near the outer edge of the Perseus Arm, lies the cluster M103. The two clusters are about the same age, but M103 is not quite as bright – it's not clear why this cluster found its way into Messier's catalogue while the more prominent NGC 457 did not. Around M103 we can see a handful of other clusters: some are associated with M103

cluster, whereas others lie nearer to us, in the main bulk of the Perseus Arm.

These clusters are old enough to have dispelled most of the gas from which they were born. But in this part of Cassiopeia, and especially towards its other border, with Cepheus, there are several clouds still forming stars today. A small telescope shows, for example, the NGC 281 cluster and nebula – a large faint triangle that John Sanford describes as 'resembling the famous North America Nebula in Cygnus minus the Florida peninsula'.

The most important object in this part of the Perseus Arm, however, is far from conspicuous to an optical telescope. It is a black, dust-filled cloud, visible to ordinary telescopes only by a small patch of light, NGC 7538, where some radiation is seeping through from newly born stars hidden deep inside.

NGC 7538 is a dark molecular cloud – which may contain as much mass as 500 000 suns – from which a small patch of light is escaping. This infrared view reveals that vigorous star-formation is taking place inside (red patches), while the blue glow is light pouring out through a 'crack' in the cloud. Hélène Dickel suggests that this is radiation from young stars in the cloud that has burst through a line of weakness, like the bubbles frothing out of an uncorked bottle of champagne.

As with other dark molecular clouds, astronomers can probe the interior of this dense region by studying infrared and radio waves that penetrate the dust. According to Hélène Dickel, of the University of Illinois, these observations reveal 'a remarkable assortment of young objects . . . The tremendous variety makes this region a key candidate for studying the process of star formation and the evolution of molecular clouds'.

The dark cloud around NGC 7538 contains as much mass as 50 000 to 500 000 suns, and at its heart is a cluster of freshly formed stars. These are heating up the molecules in the vicinity, prompting some of them to pump out intense radio waves as a maser, the microwave equivalent of a laser. Astronomers have found maser emission from hydroxyl (OH) and water (H_2O) here – and it was the first place in the Galaxy where they found similar emission from formaldehyde molecules (H_2CO).

Dickel suggests that the powerful radiation from the stars in the dense cloud has burst out through a line of weakness, to force open a passage from the centre to the outside of the cloud, at the position of the optical nebula NGC 7538. This she likens to the eruption from a champagne bottle when the cork is drawn: the nebula we see with an optical telescope is the froth from a cosmic bottle of bubbly.

If we could look at this part of the sky with eyes sensitive to radio waves instead of light, this star-forming cloud would be overshadowed by a far more brilliant neighbour: Cassiopeia A. As observed from Earth, Cassiopeia A is the strongest radio source in the sky – and it is certainly one of the most powerful individual radio emitters in the entire Galaxy.

Cassiopeia A is brilliant at infrared and X-ray wavelengths, as well as radio. But to the optical astronomer, it is a disappointment: a straggle of extremely faint filaments. These optical observations have, however, proved critical in understanding this object, because they show that its gases are speeding outwards at a rate of 6 000 kilometres per second. Only the explosion of a supernova could fling out gases so fast – and could provide the tremendous energy needed to keep this object shining at infrared, X-ray and radio wavelengths.

BOX 2. **How the faintest supernova gave birth to the brightest radio source**

If our eyes were sensitive to radio waves instead of light, the sky would look very different. Apart from the Milky Way and the Sun, the most brilliant radio-emitting object – the 'Sirius' of the radio skies – is Cassiopeia A.

The first celestial radio source to be identified was the Galaxy itself, discovered by Karl Jansky in the 1930s (see Chapter 3). This work was followed up by Grote Reber, an amateur astronomer and radio ham, during the Second World War. Reber published the first maps of the radio sky, and we can clearly see Cassiopeia A as a 'bump' on his contour maps. At first Reber thought he was looking at a spiral arm, seen end-on, but after more detailed observations he wrote: 'at +60° declination considerable energy was found near right ascension 23h 15m . . . in the region of Cassiopeia'.

Unfortunately, Reber could not confirm that there was an individual object in this region of sky because 'winter was closing in and the equipment was scheduled to be moved . . . [so] only a few exploratory traces were secured'. As a result, the 'official' discovery was made by professional radio astronomers at Cambridge a few years later.

At first, optical astronomers searched in vain for an object that could be emitting these radio waves. In 1951, Graham Smith at Cambridge pinned down the position of the source with high precision. This gave optical astronomers in California the confidence to spend precious time searching with the newly opened '200-inch' Hale Telescope on Palomar Mountain.

Using the giant telescope, Walter Baade and Rudolph Minkowski managed to photograph what they called 'a galactic nebulosity of extraordinary properties'. It was a scatter of very red filaments, with spectra showing that they had very high speeds – thousands of kilometres per second in some cases. Iosef Shklovskii in the Soviet Union immediately suggested that these gases came from an old supernova; but Baade and Minkowski did not accept this interpretation at first. Indeed, they stated 'there is every reason to believe that the Cassiopeia source has nothing to do with supernovae'.

Over the next few years, however, astronomers found that the filaments were moving across the sky, obviously expanding rapidly from a central point. By backtracking the motions of the gas, they could say when the supernova exploded. Canadian astronomer Sidney van den Bergh has made the most exhaustive study of all the data, and concludes that we should have seen the star explode around the year 1660, give or take a few years.

What is most odd is that no-one recorded a supernova at that time. Cassiopeia is visible from Europe at any season and any time of night, and the late seventeenth century was an active time for observers: Halley, for example, saw 'his' comet in 1682.

Was the supernova just too far away? The answer must be 'no'. The expansion of the Cassiopeia A 'supernova remnant' gives us a fairly precise way to measure its distance, and it turns out to lie just over 9 000 light years away: firmly in the Perseus Arm. If the supernova were similar to those we see in other galaxies today, it should have been bright enough to be seen in the daytime. Even if it had resembled the underluminous SN1987A in the Large Magellanic Cloud, it would have outshone the brightest stars at night.

In 1980, William Ashworth, an American historian of astronomy, took one step closer to solving the puzzle. He discovered that England's first Astronomer Royal, John Flamsteed, actually recorded a star near the position of Cassiopeia A in August 1680, and named it '3 Cassiopeiae'. Later astronomers, however, found that this star did not exist, and deleted it from subsequent catalogues.

Ashworth thinks that Flamsteed's 3 Cassiopeiae probably was the supernova that produced Cassiopeia A, even though the positions do not tally exactly. This discrepancy was probably the result of errors in Flamsteed's measurements, according to Ashworth. He buttresses his argument by asking: 'If Flamsteed was not observing Cassiopeia A on the night of 16 August 1680, just what was he looking at?'

Flamsteed's star was at magnitude 6, just at the limit of naked-eye visibility. The supernova probably did not get much brighter than this, or many other people would have noticed it too. This means the supernova would have been only 250 000 times brighter than the Sun: only one-thousandth as brilliant as the faint 1987 supernova in the Large Magellanic Cloud, and not much more luminous than the brightest ordinary stars in our Galaxy – for example, phi Cassiopeiae, which also lies in the Perseus Arm.

What is still not clear is how a star could have exploded in such a modest manner, and yet give rise to one of the most powerful radio sources in the entire Galaxy.

With modern radio telescopes and X-ray telescopes, we can observe the structure of Cassiopeia A in fine detail. And what we see certainly confirms our notion that we are observing the aftermath of an immense explosion. The images show an irregular ball of energy, resembling photographs of atom bomb tests that have been taken just a fraction of a second after ignition. And, on the cosmic scale, the three centuries that have elapsed since this supernova erupted are indeed just an instant of time.

The supernova that produced Cassiopeia A must have been unusually dim, because astronomers did not see it explode, but there's nothing meagre about the wreck we see now. The exploding star threw off some 20 solar masses of gas, and this material is now cannoning into a shell of debris that the star lost at an earlier stage, when it was an unstable red giant. The

FACING PAGE Aftermath of an explosion: this colour-coded radio image of Cassiopeia A reveals the tangled wreck of expanding gases from a star that must have blown up about 300 years ago – but that no-one saw explode. The colours represent the intensity of the radio emission, with red the most intense, and blue the weakest. The emission comes from electrons moving in strong magnetic fields.

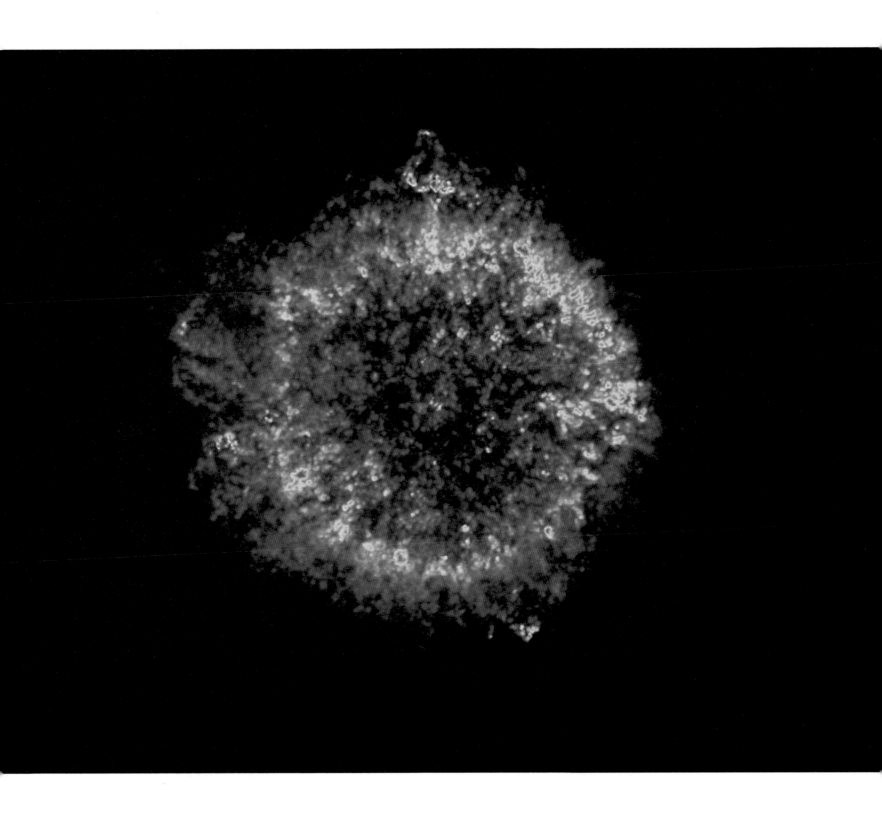

result is a tangle of magnetic field, where high-speed electrons are accelerated and trapped, and finally lose their energy by emitting powerful radio waves.

The Perseus Arm is, in fact, quite a graveyard of stars. A radio telescope reveals at least nine supernova remnants strung along the segment that we show in the map, which reaches from the constellation of Cepheus to Monoceros. Some are quite large and dim, even when seen at radio wavelengths, and these are probably thousands of years old. But there are another three remnants of stars that exploded recently enough to have been recorded in the historical annals.

Making our way back from Cassiopeia's boundary with Cepheus towards Perseus, we come upon the remains of one of the most famous stars in history: the 'nova' of 1572. This star was extensively studied by the great Danish astronomer Tycho Brahe, and so we know quite a lot about it. Tycho measured its changing brightness so accurately that we can tell what kind of a supernova it was: a Type I – the explosion of a dwarf star that collapsed under its own weight when a companion star piled matter onto it.

The exploding star here was of a quite different type from the supernova that gave rise to the neighbouring Cassiopeia A remnant. The latter must have been a large heavyweight star that exploded without producing much light. Tycho's super-nova was a small star, little heavier than the Sun and no larger than the Earth, which exploded in a burst of radioactive energy that made it shine more brightly than a billion suns.

But the remains of the supernovae, after three or four centuries, look very similar. The reason is a matter of simple physics. In each case, we have the explosion of something fairly small (compared with the size of the remnants now) into a large region of surrounding gas. The gas from the exploding star sweeps up the interstellar gas, into a dense expanding shell. After a couple of centuries, this shell has largely forgotten the exact kind of explosion that produced it. All that's important is the time that has elapsed, the total energy of the original explosion and the density of the gas that it is expanding into.

Mathematically, in fact, a supernova is exactly the same as the explosion of an atomic bomb on the Earth. A nuclear weapon is, again, the eruption of a small but powerful device into a surrounding medium of gas: the atmosphere. The equations that astronomers use to describe supernova remnants

Danish nobleman, bon viveur and meticulous observer, Tycho Brahe was the most accomplished astronomer of his time. His measurements of the movements of the planets provided his assistant Kepler with proof that they circled the Sun – and not the Earth – in elliptical paths. But his chance observation of the supernova of 1572 – which proved that the 'unchanging' heavens changed – stimulated others to start exploring the sky after centuries of scientific suppression. In a sense, Tycho's super-nova started the scientific revolution that is still in full spate today.

Optical telescopes show very little at the spot where Tycho's supernova exploded (in the constellation of Cassiopeia). At 40 million degrees Celscius, the scattered remains are still too hot to produce light – but they do shine brilliantly in radio waves and, as shown here, X-rays. This colour-coded image was recorded by the Exosat satellite, with the most intense X-radiation coming from the regions coloured red. The slightly dented outline is caused by the shell running into smaller, denser clouds of interstellar gas that have slowed its expansion. But expansion still continues apace: in the four centuries since the explosion, the remnant has grown to a size of 17 light years.

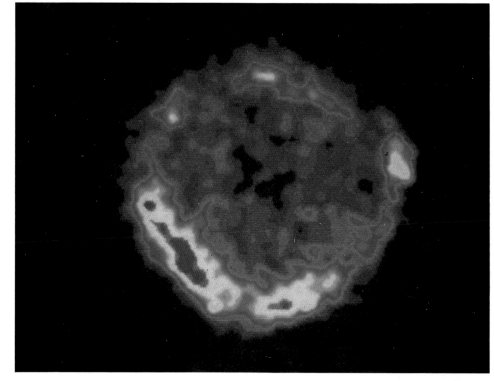

BOX 3. The star that changed the world

On the evening of 11 November 1572, the leading astronomer of the time was taking a walk before dinner, when 'directly overhead, a certain strange star was suddenly seen, flashing its light with a radiant gleam. Amazed, and as if astonished and stupified, I stood still, gazing for a certain length of time with my eyes fixed intently upon it and noticing that same star placed close to the stars which antiquity attributed to Cassiopeia'.

Tycho Brahe had every reason to be taken aback. The scientific teaching of the time was the philosophy of Aristotle, as backed by the Church. Its best-known tenet in astronomy was that the Earth was the centre of the Universe, and everything else moves around the fixed Earth.

Just as central, however, was the idea that things become more perfect the further they are from the Earth and the nearer to Heaven. Corruption and decay only occurred on the Earth and in its atmosphere: further out, the Moon was mildly imperfect – it has dark patches, and changes its shape during the month. Beyond the Moon lay the planets, which were perfect in themselves, and changed their positions in a well-ordered way. Finally there was the sphere of the stars, closest to Heaven. Apart from their stately wheeling, night after night, the stars were absolutely perfect and unchanging.

So Tycho was completely thrown: 'When I had satisfied myself that no star of that kind had ever shone forth before, I was led into such perplexity by the unbelievability of the thing that I began to doubt the faith of my own eyes'. He first asked the servants with him if they too saw the star. They confirmed that there was a bright star – but a nobleman could never tell if his servants were merely saying what was expected of them. Only when some people passing in carriages confirmed the star's existence, did Tycho believe the evidence of his own eyes.

The object in itself might not contradict the Universe of Aristotle. Astronomers were well aware of comets, for example, but thought that they occurred in the Earth's atmosphere. Tycho immediately decided to see if the star showed any sign of moving. 'Immediately I got ready my instrument. I began to measure its situation and distance from the neighbouring stars of Cassiopeia'.

This object, however, showed every sign of being a bone fide star – as the months went by, it stayed in exactly the same place. Tycho concluded that it 'is located neither . . . below the Moon, nor in the orbits of the seven wandering stars [planets] but in the eighth sphere, among the other fixed stars'.

The new star was the first visible sign that the old theory was wrong. The Polish monk Nicolaus Copernicus had already suggested that the Earth might move around the Sun, but there was no hard evidence to back this hypothesis against the prevailing ideas.

Fortuitously, a bright comet appeared only five years later. Again, Tycho measured the position of the comet with unprecedented accuracy, and discovered that it, too, lay well above the atmosphere. Indeed, the comet's path would have taken it crashing through the crystalline spheres that were supposed to move the planets along.

The full impact of his measurements of the new star and the comet was, ironically, lost on Tycho himself. He continued to believe that the Earth was the centre of the Universe, although he made the other planets orbit the Sun, which was still going around the Earth. But other scholars of his day worked through the logical conclusions of his observations.

Because Tycho's results destroyed the distinction between the mutable material of the Earth and the unchanging matter of the heavens, there was every reason to accept the idea that the Sun lay at the centre of things – especially when Tycho's assistant, Johannes Kepler, showed that he could explain the planets' motions simply in terms of ellipses with the Sun at one focus.

If the planets were not moved by crystalline spheres, then why should the stars be all attached to a sphere? Thomas Digges, in England, drew a diagram of the planets (and Earth) orbiting the Sun, with the stars scattered beyond, with the words: 'This orb of stars . . . infinitely up extendeth itself.'

The momentum became unstoppable. The Dominican monk Giordano Bruno believed that 'there are countless suns and an infinity of planets which circle round their suns'. Another supernova appeared in 1604, as if to mock those who still clung to the old ideas. The young Galileo was brought up in this new atmosphere: when he turned the telescope to the sky, he was largely confirming what he expected to see: imperfections on the Moon and Sun, the phases of Venus as it circled the Sun, and stars without number.

Within a lifetime, the claustrophobic Universe of Aristotle was swept away, and replaced with the vast three-dimensional Cosmos we know today, where, at last, we have been able to place Tycho's star in its true perspective – at a distance far greater than any mind of that time could have envisaged.

were in fact first calculated in the late 1940s by physicists trying to describe the blast from nuclear bombs.

According to the theory of Type I supernovae, the original star should blow itself apart entirely, unlike a Type II supernovae, which should leave a tiny, compressed core – usually a neutron star, an object only 20 kilometres across but extremely dense. A neutron star only a few centuries old would be still be very hot – at a temperature of millions of degrees – and it would shine so brightly at X-ray wavelengths that we would see it clearly in X-ray images of Tycho's supernova remnant. X-ray observations confirm the theory of Type I supernovae, by showing us that the interior is indeed empty.

Not far from Tycho's remnant in the sky, however, we find a remnant that is very different. The unexcitingly named 3C 58 – from its place in the third Cambridge catalogue of radio sources – was something of an oddment until the 1970s. At radio wavelengths, it has an unusual egg-shape, and astronomers were even uncertain whether it was an object in our Galaxy or a radio galaxy far beyond.

A British astronomer, Richard Stephenson, eventually discovered its true nature. Checking through ancient oriental records of the sky, he found a 'guest star' that appeared in the

With its red filaments looking like grasping pincers, the Crab Nebula certainly lives up to its nickname – unlike most other supernova remnants (which tend to be spherical in shape). The Crab is the remains of a star that exploded in AD 1054, and whose gases are still moving outwards at about 1500 kilometres per second. The eerie blue glow inside the filaments is synchrotron radiation – the first to be detected occurring naturally in the Universe. It comes from energetic electrons ejected from the Crab Nebula's central pulsar.

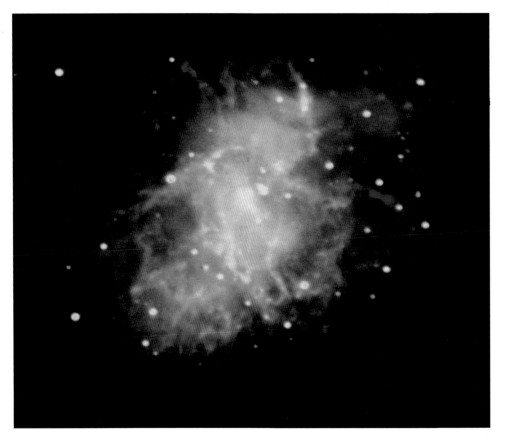

constellation Cassiopeia in 1181. 'It was like Saturn and its colour was bluish-red and it had rays', reads one account. According to another, the astrological interpretation was 'at any moment we can expect control of the administration to be lost'.

Stephenson unravelled the Chinese descriptions of where the supernova lay in the sky. In particular, the official history of the Sung Dynasty states that the guest star 'invaded' the Chinese constellation Ch'uan-shê. Meaning 'The Inns', this constellation is a straggling line of stars above the W that western tradition calls Cassiopeia. Right on this roadway of inns is the radio source 3C 58.

The latest measurements of the distance to 3C 58 confirm Stephenson's suggestion by putting it firmly within our Galaxy, right in the middle of the Perseus Arm. And its appearance at radio wavelengths turns out not to be so unusual: it

resembles perhaps the most famous single object in the Perseus Arm, the Crab Nebula.

To reach the Crab, we continue our journey back along the Arm, past the Perseus Double Cluster where we began this chapter, and on into the region of this spiral arm that lies behind the constellation of Taurus. Just above one 'horn' of the Bull we find the Crab Nebula – an object with a special place in the development of the 'new astronomy' in the 1950s and 1960s. At that time, a leading British astronomer, Geoffrey Burbidge, even remarked 'there are two kinds of astrophysics: the astrophysics of the Crab Nebula and the astrophysics of everything else!'

If we could view the Perseus Arm from outside the Galaxy, then the Crab Nebula would put on a fairly poor show, as compared with the star clusters and regions of star-formation. But that applies only to the optical view. At radio wavelengths, the Crab would be a brilliant object, second only to Cassiopeia A. And viewed in X-rays or gamma rays, the Crab is the jewel of the Perseus Arm.

As a result, astronomers have 're-discovered' the Crab Nebula every time they have started to scan the skies at a new wavelength. The first of these 'new astronomies' was radio astronomy. The first radio sources (outside the Solar System) to be discovered were Cassiopeia A and Cygnus A – but these were not identified until many years later. The Crab Nebula was in the first batch of three radio sources that were linked to known objects. In the 1960s, it was the second X-ray source to be detected, and a decade later satellites showed it is the second brightest gamma-ray source in the sky, as seen from the Earth.

What is unique about the Crab Nebula is that it is so brilliant across the whole spectrum of wavelengths. By observing it at wavelengths from radio to gamma rays, astrophysicists have been able to use it as a natural laboratory for high-energy experiments way beyond our means on the Earth.

The Crab is not a 'nebula' in the ordinary sense of the word: a region of star-birth. Indeed, it is the opposite: the tattered gases from a star that has died. The Chinese recorded the supernova that created the Crab Nebula in the year 1054. It was a 'guest star' so brilliant that it was visible in daylight for three weeks, and at night for nearly two years.

This star was unknown to western astronomers when, in 1731, an English astronomer, John Bevis, came across a misty patch of light in this position. The great comet-hunter Charles Messier stumbled over the nebula in 1758, as he was trying to relocate a comet seen a few weeks earlier. This nebula so resembled a tail-less comet that it inspired Messier to draw up his famous catalogue of fuzzy objects that could be confused with comets. He recorded it as 'a whitish light elongated like the light of a candle'.

A century later, the third Earl of Rosse observed the nebula with his colossal and unwieldy telescope in central Ireland. He discerned 'resolvable filaments . . . springing principally from its southern extremity'. His first drawing resembles a pineapple more than anything else, but by the mid-1850s the Earl was seeing the nebula as an oval with a gap at one end – like the claw of a crab. This appearance probably led him to the nickname the 'Crab' Nebula.

In the first decades of the twentieth century, several astron-omers in the United States realized that the nebula is expanding rapidly, either by studying the Doppler shift of lines in its spectrum or by comparing photographs taken several years apart. They calculated that the expansion must have begun about 900 years earlier, and made the identification with the supernova of 1054. This gave the Crab Nebula the distinction of being the first 'supernova remnant' to be identified.

The supernova remnants detected subsequently have mainly been rather faint at optical wavelengths. The Crab is anomalously bright. As seen from the Earth, it appears at magnitude 8, so you can just make it out with binoculars if the sky is really dark. Photographs show a smooth bluish nebula, surrounded by a filaments of reddish gas.

In the 1950s, Soviet astronomers made the remarkable discovery that the bluish light is not coming from hot glowing gases. Instead, the light is generated by fast electrons whirling in a magnetic field that fills the nebula. Physicists had already observed such radiation from synchrotron particle accelerators in the laboratory, but this was the first proof that 'synchrotron radiation' occurs naturally in the Universe. The discovery confirmed the prediction by Iosef Shklovskii that synchrotron radiation accounts for many of the powerful sources of radio waves in the sky.

Until the late 1960s, however, it was not clear why a remnant 900 years old should be quite so active. Other supernova remnants, to be sure, produce powerful radio waves and some X-radiation as their high-speed ejecta crashes into the interstellar gas. But the Crab is brightest at its centre, not near its periphery.

Some astronomers had already suspected that an unusually blue star in the centre of the nebula might, somehow, be responsible for its brightness, and that this star might be the collapsed core of the giant star that had exploded as the supernova of 1054. Confirmation came, however, in a roundabout way.

In 1967, researchers at Cambridge discovered four 'pulsing' radio sources. These 'pulsars' turned out to be rapidly spinning neutron stars, the condensed cores of old supernovae. We see a pulse of radiation each time the star rotates, like seeing 'flashes' from a rotating lighthouse lantern. These explosions occurred so long ago that the gaseous supernova remnants have long since dissipated into space – but might there be

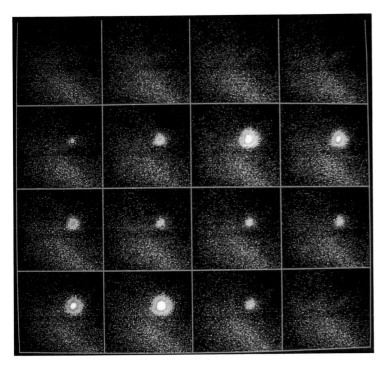

The pulsar at the centre of the Crab Nebula is responsible for many of the nebula's bizarre properties. It spins 30 times a second and – unlike most pulsars – emits optical flashes as well as those at radio and X-ray wavelengths. This false-colour sequence of X-ray flashes was obtained by taking sixteen successive frames over one rotation period. The frames are separated by two-thousands of a second and reveal two main flashes per rotation. This must mean that the Crab Pulsar has two hot spots, or beams of radiation, on (or just above) its surface, rather than just one.

pulsars inside the remnants of the supernovae observed in historic times?

Dave Staelin and Ed Reifenstein in the United States investigated several of the known supernova remnants – and found pulses coming from the Crab. What was surprising was the repetition rate: 30 times per second, far faster than any other pulsar known then, and clearly a sign of the pulsar's youth. Knowing this period, optical astronomers were then able to pick up regular flashes of light coming from the 'blue star' at the nebula's heart.

Astronomers are now certain that the pulsar is responsible for most of the Crab's unusual properties. Its rotation is gradu-

ally slowing down, and the energy it loses is converted into the powerful magnetic fields and fast electrons that make the nebula so brilliant right across the spectrum.

For many years, astronomers thought the Crab Nebula was unique. They have now found an almost identical nebula, complete with pulsar spinning 20 times per second, in our neighbouring galaxy, the Large Magellanic Cloud. And there are similar, but less flamboyant examples in our Galaxy – for example, 3C 58, which we passed earlier in this chapter. But the Crab Nebula is still the nearest test-bed for many high-energy processes that astronomers need to understand when they come to deal with the much more powerful radio galaxies and quasars that we find far out in space.

Not far from the Crab is a very different looking supernova remnant, IC 443. Here we see bright filaments around the edge, where the supernova's gases have swept up interstellar material, and an empty centre. Invisible to an ordinary telescope is a dark cloud, rich in molecules, that runs across the middle of IC 443. Where the supernova remnant has run into the cloud, it has swept up and compressed about 1000 solar masses of gas. In protest, the hydrogen molecules here are radiating away much of this energy in the form of spectral lines at infrared wavelengths. If we could see these wavelengths, we would find IC 443 shining as brightly as 1600 suns – one of the most luminous sources of its type in the Galaxy.

The brief history of this region is that a massive star formed inside the molecular cloud, and eventually – some 3000 years ago – exploded as a supernova. The gases from this supernova are now compressing the remaining cloud material – except to the north-east and south-west, where the gases have broken through the cloud and we see the optical filaments.

This part of the Perseus Arm contains other signs of recent star-formation. IC 443 lies in an association of young stars, Gemini OB1, which is one of two large groupings of young stars in this region. The second is Auriga OB1. This association of young stars contains two prominent star clusters, with an even more splendid example at its edge.

Even a pair of binoculars will reveal these three star clusters lying within the constellation Auriga, and Messier included them in his catalogue as numbers 36, 37 and 38. They are among the finest sights in the northern sky for a small telescope. The stars of M36 and M38, the two clusters in the Auriga OB1 Association, are each arranged in a square or cross-shaped pattern. M36 – with an age of only 20 million years – is the younger and more concentrated of the two.

But the most magnificent of the three is M37. Admiral W. H. Smyth described it as 'a magnificent object; the whole field being strewn, as it were, with sparkling gold dust'. According to another Victorian astronomer, the Revd T. W. Webb, it is 'extremely beautiful . . . gaze at it well and long!' This cluster is rather older, having formed about 300 million years ago, and some of its stars have now turned to red giants. This gives it a rather different appearance from M36 and M38, which are dominated by brilliant blue stars.

In the constellation Monoceros, we come across an even younger, and more brilliant, cluster of stars. It is another cluster overlooked by Messier, even though the first English Astronomer Royal, John Flamsteed, had described it decades earlier. As a result, it now bears the rather unglamorous name of NGC 2244: but it is better known as the cluster at the heart of the Rosette Nebula.

You can spot the star cluster NGC 2244 with the naked eye, on a very clear night, and through a telescope it looks like a small heap of blue-white diamonds. These stars form a stunning contrast to 12 Monocerotis, a yellow star that appears to be the brightest in the cluster, but in fact merely happens to lie in front of NGC 2244, at only one-tenth its distance. But the famous Rosette Nebula is a disappointment to anyone 'eyeballing' through a telescope. Even on a clear night, with a moderate-sized telescope, it appears only as 'a large, circular shell of faint grey nebulosity', in the words of David Eicher, an American veteran deep-sky observer.

Long-exposure photographs, however, show the Rosette as one of the glories of the northern sky. Around the blue-white stars at its core, the petals glow red with the light from hot hydrogen. The Rosette is tremendous not only in its looks but in its size. Even at a distance of 5500 light years, it appears twice the size of the Moon in our skies – meaning that it is about a hundred light years across. This is over five times larger than the more famous Orion Nebula: or, to put it in context, the entire Orion Nebula would fit into the hole in the centre of the Rosette!

The central stars of the Rosette formed about a million years ago, and their intense radiation and powerful winds have blown away the gas that initially cocooned them. Having cleared a space immediately around them, the stars now illumi-

nate the large shell of dispersing gases that form the Rosette. On photographs of this nebula, we can see small dark spots within the petals of the Rosette. These are regions of denser gas and dust in the original cloud that have been laid bare as the less dense regions have been eroded by the energy from the central star cluster.

The Rosette lies in an active region of star-formation and young stars. It is at the edge of a dark molecular cloud, where stars have yet to form, and nearby is a supernova remnant, the Monoceros Loop, which is probably the remains of a massive star that lived a profligate life and exploded young. These nebulae and clouds are surrounded by an association of young stars, Monoceros OB2. Among these is an object with a special claim to fame: Plaskett's Star, a pair of heavyweight twins with the highest masses of any stars to have had their weights measured directly.

Plaskett's 'Star' is in fact a very close double star – a circumstance that allows astronomers to weigh the pair directly by using Newton's law of gravitation. In 1922, the Canadian astronomer J. S. Plaskett studied this particular star in Monoceros – at magnitude 6, just visible to the naked eye – and found that it consisted of two nearly matched stars, orbiting one another every 14 days. According to Plaskett's measurements, each was 55 times heavier than the Sun – far heavier than any other star known.

Since then, Plaskett's Star has kept its heavyweight title (although astronomers suspect, from indirect methods, that some stars may be even more massive). The most recent results suggest that the stars in this massive pair weigh in at 43 and 51 suns.

Although we have described IC 443, the Auriga clusters and the Rosette Nebula as lying in the Perseus Arm, the true

FACING PAGE The glorious Rosette Nebula in Monoceros surrounds the cluster NGC 2244, whose stars were born out of it. Five times larger than the famous Orion Nebula – which would fit into the hole in the middle – the Rosette lies 5500 light years away, and yet appears bigger than the Full Moon in our skies. The stars in NGC 2244 were formed about a million years ago, and their powerful radiation and strong stellar winds are blowing away the gas that once shrouded them. The whole region is part of a large complex of star-formation, situated at the edge of a molecular cloud whose stars are yet to be born.

situation seems to be more complicated. Indeed many astronomers would say they lie at the outer edge of the Orion Arm – even though there is a clear, if narrow, gap in between.

The problem is that the neat and tidy 'textbook' picture of the Perseus Arm breaks down here. In the stretch of arm between the Double Cluster and the Crab Nebula, we see very few of the nebulae and young star clusters that we normally use to delineate a spiral arm. Radio telescopes tuned to a wavelength of 21 centimetres, however, do pick up emission from interstellar hydrogen in this region. If it were not for this radiation, we might think that the Perseus Arm disappeared altogether in the constellation that gives it its name!

Once we come to the region of the Crab Nebula and IC 443, we begin to pick up nebulae and young star clusters again. But the 'spiral arms' they trace do more to confuse than to clarify the picture. They do not continue the line of the Perseus Arm that we have followed before. Indeed, they seem to define two different arms.

If we follow a direction towards the Crab Nebula, our path is waymarked by a dense molecular cloud (associated with a nebula called S235) and more faint nebulae. These mark out an 'arm' that heads steeply down towards the bottom right of our map and beyond, at an angle of 45 degrees to the direction of the galactic centre. This seems much too steep for a major arm.

We could, instead, follow a number of other faint nebulae on an inner path through the 'gap' in Perseus, bringing us towards the supernova remnant IC 443. This possible arm is marked by the young clusters in Auriga, by the molecular cloud around IC 443 and the region of the Rosette Nebula. The problem with this path is that it is not steep enough: the young objects in Auriga, Gemini and Monoceros lie at about the same distance from the Galaxy's centre as do the Double Cluster and the star clusters and nebulae in Cassiopeia. If this

is a 'spiral arm', it is forming an arc of a circle, and is not spiralling outwards at all!

The 21-centimetre radio observations reveal hydrogen in these two 'arms', extending further on round the Galaxy in each case. In principle, we could see which of these two hydrogen arms connects with the Perseus Arm in the region of Perseus itself. Unfortunately, nature has conspired to make this task almost impossible. The connection is made in a region that happens to lie, as seen from the Earth, in the opposite direction to the Galaxy's centre. The gas here travels exactly across our line of sight, without any speed towards or away from us. And without any speed to measure, we cannot use the usual method for telling how far away a cloud of hydrogen lies.

The truth is probably that astronomers have generally been too simple-minded about the whole thing: they have wanted a continuous, smooth long Perseus Arm, and have been trying to straitjacket the observations to fit. The observations suggest that the Perseus Arm is more complicated. It is a continuous arm from beyond Cassiopeia A up to the Double Cluster, but in the region of Auriga and Taurus it either branches into two – something we often see in other spiral galaxies – or it peters out into hydrogen gas, and our telescopes then pick up the beginnings of two more segments of arms that start at different distances from the galactic centre, in the vicinities of IC 443 and the Crab Nebula respectively.

This detailed look confirms what we found in Chapter 3, when a plot of spiral arm tracers over the whole Galaxy showed that the 'Perseus Arm' seems to consist of discontinuous patches of young stars and nebulae – like the 'shingles' we see in many other spiral galaxies. Unlike the Sagittarius Arm, which is a major arm wrapping itself all around the Galaxy, the Perseus Arm is not part of a 'grand design' spiral. Nonetheless, it is a treasurehouse of rare and beautiful sights for astronomers.

CHAPTER 5

The Orion Arm

A

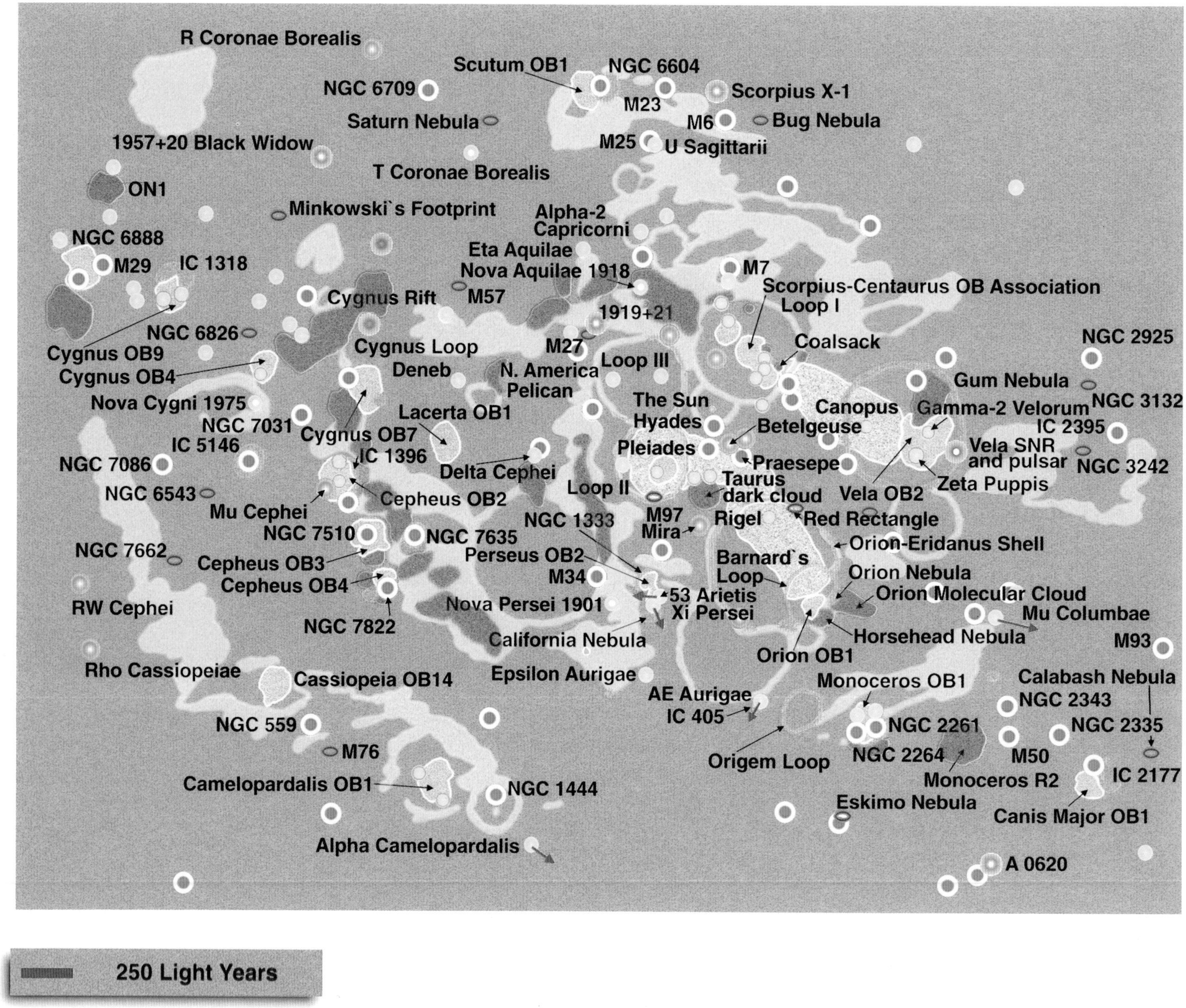
R Coronae Borealis
Scutum OB1
NGC 6604
NGC 6709
Scorpius X-1
M23
Saturn Nebula
M6
Bug Nebula
1957+20 Black Widow
M25
U Sagittarii
T Coronae Borealis
ON1
Minkowski`s Footprint
Alpha-2
Capricorni
NGC 6888
Eta Aquilae
M29
IC 1318
Nova Aquilae 1918
M7
Scorpius-Centaurus OB Association
Cygnus Rift
M57
Loop I
1919+21
NGC 6826
Coalsack
M27
NGC 2925
Cygnus OB9
Cygnus Loop
Loop III
Cygnus OB4
Deneb
N. America
Gum Nebula
Pelican
Nova Cygni 1975
The Sun
NGC 3132
Lacerta OB1
Canopus
Gamma-2 Velorum
NGC 7031
Hyades
Betelgeuse
IC 2395
IC 5146
Cygnus OB7
Pleiades
Vela SNR
and pulsar
IC 1396
NGC 7086
Praesepe
NGC 3242
Delta Cephei
Taurus
dark cloud
Zeta Puppis
NGC 6543
Loop II
Cepheus OB2
Vela OB2
Mu Cephei
M97
Rigel
Red Rectangle
NGC 1333
Mira
NGC 7510
NGC 7635
Orion-Eridanus Shell
NGC 7662
Perseus OB2
Barnard`s
Loop
Cepheus OB3
Orion Nebula
M34
Cepheus OB4
53 Arietis
Orion Molecular Cloud
RW Cephei
Nova Persei 1901
Xi Persei
Mu Columbae
NGC 7822
California Nebula
Horsehead Nebula
M93
Orion OB1
Rho Cassiopeiae
Epsilon Aurigae
Cassiopeia OB14
Calabash Nebula
AE Aurigae
Monoceros OB1
NGC 2343
IC 405
NGC 559
NGC 2261
NGC 2335
M76
Origem Loop
NGC 2264
M50
Camelopardalis OB1
IC 2177
NGC 1444
Monoceros R2
Eskimo Nebula
Canis Major OB1
Alpha Camelopardalis
A 0620
250 Light Years

B

KEY TO MAP A OVERLEAF

KEY TO MAP A

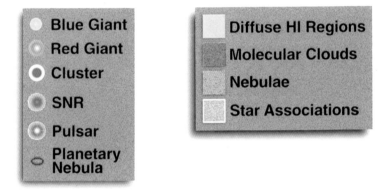

- ● Blue Giant
- ◉ Red Giant
- ○ Cluster
- ◎ SNR
- ◉ Pulsar
- ⊖ Planetary Nebula

- ☐ Diffuse HI Regions
- ■ Molecular Clouds
- ☐ Nebulae
- ☐ Star Associations

MAP A In the Local Arm, we can discern a lot of structure in the interstellar medium. Regions of enhanced density (yellow) have been mapped by studying the amount that their dust particles absorb light coming from more distant stars. Molecular clouds are located by both their emission of radio waves and their heavy absorption of light.

MAP B Swirls of gas and dust mark the Local Arm as it heads from the region of Cygnus (left) to Vela (right): dense dark molecular clouds form an inner rim to the Local Arm. The Orion region (lower right) is a spur stretching out towards the Perseus Arm. From the side, we would see that many of the young stars and their associated molecular clouds and nebulae form a tilted 'flipper' – Gould's Belt – that dips into the page in the Orion region.

Striding high in the skies of the northern winter is the mighty hunter, Orion – one of the most brilliant constellations of all. Orion is among the few constellations with a shape that bears some resemblance to its name. Four stars clearly mark his shoulders and knees, while a line of three in the middle delineates his belt. A triangle of faint stars shows us where his head lies; and a curve of stars, towards Taurus, outlines the shield that protects him from the charging bull.

In legend, Orion was the greatest of hunters – but brought down by his own pride. According to one legend, he boasted that he could slay all wild animals. Determined to punish Orion's chutzpah, Hera made a scorpion rise from the ground and administer a fatal sting to the heel of the giant hunter. The gods then raised both Orion and the scorpion to the heavens – but positioned safely on opposite sides of the sky. So, when Scorpius rises, Orion sets; and when the hunter comes up above the horizon, the scorpion slinks back beneath the ground.

Colourful tales aside, the constellations of Orion and Scorpius do have an important astrophysical link. The main stars in these two regions of the sky are among the youngest and brightest stars to have formed in our part of the Galaxy – in the spiral arm where the Sun currently lies. In terms of scientific nomenclature, the scorpion has lost this particular battle, because astronomers refer to this region as either the Local Arm or the Orion Arm.

In fact, astronomers could have named our own arm after virtually any constellation: all the bright stars that delineate the constellation shapes lie in the Local Arm. Because we live within the Local Arm, we have a worm's eye view: it stretches away from us in two directions on opposite sides of the sky, forming regions where we see stars, nebulae and dark clouds at a variety of distances all piled up behind one another. The arm comes outwards from the direction of Cygnus, and then passes us as it heads towards Vela. Both these constellations are rich in bright young stars – but they are also choked with dust and dark molecular clouds.

Orion is rather offset from the main course of our spiral arm, but it does have a special claim. This region is an active nursery, where brilliant new stars are being born now. If we were flying into the Galaxy from the outside, our attention would hardly be drawn by the dark clouds of Cygnus and Vela or the extended star-associations of Scorpius. We

would home in on the brilliant young stars, and protostars, of Orion.

The young stars of Orion, with the accompanying nebulae, lie about 1600 light years from the Sun. Beyond them, the outermost limits of the Orion Arm are found in the constellations of Monoceros, Canis Major and Puppis, several thousand light years away. In the early days of mapping the Galaxy, it was believed that our Local Arm stretched from Cygnus to Orion, instead of Vela. As a result, some astronomers suggested that our Local Arm should be demoted to the 'Orion Spur', signifying a mere appendage of the Perseus Arm. But the most recent surveys of molecular clouds in Cygnus and Vela show that the Orion Arm is in fact every bit as massive as the Perseus Arm.

Canis Major – the great dog – is marked out for us by Sirius, the brightest star in our skies. But Sirius is prominent only because it is one of the closest stars to the Sun. Near it in the sky, but nearly a thousand times further away, is the most interesting object in the constellation: Canis Major R1. It has been a vital clue to the way in which dying stars can trigger the birth of new stars.

Like many objects in the Galaxy, this region of stars and nebulae has several different names, depending on what you are looking at. The loose gaggle of young stars, Canis Major OB1, contains at least one cluster of stars that was first noticed by William Herschel in the eighteenth century.

A century later, the pioneer astrophotographer Isaac Roberts took a plate of this region and found, to his surprise, an enormous but very faint nebula. It is more than two degrees across – four times the apparent size of the Moon. This nebula is officially known only by its entry in the Index Catalogue compiled by the Danish astronomer Johann Dreyer in 1908, as IC 2177. Its curved symmetrical shape, however, automatically brings to mind the idea of a flying bird. Some astronomers have called it the Eagle Nebula (risking confusion with M16 in Serpens) or the Seagull Nebula. Perhaps the most fitting name is that chosen by the Revd Ronald Royer, one of today's leading amateur astrophotographers: the Dove Nebula. Whatever its avian type, it is certainly a large bird, with wings spanning over 100 light years.

Interest in the Dove Nebula heightened in the mid-1960s, when Sidney van den Bergh and René Racine, in Canada, surveyed the sky for small patches of illuminated dust that are

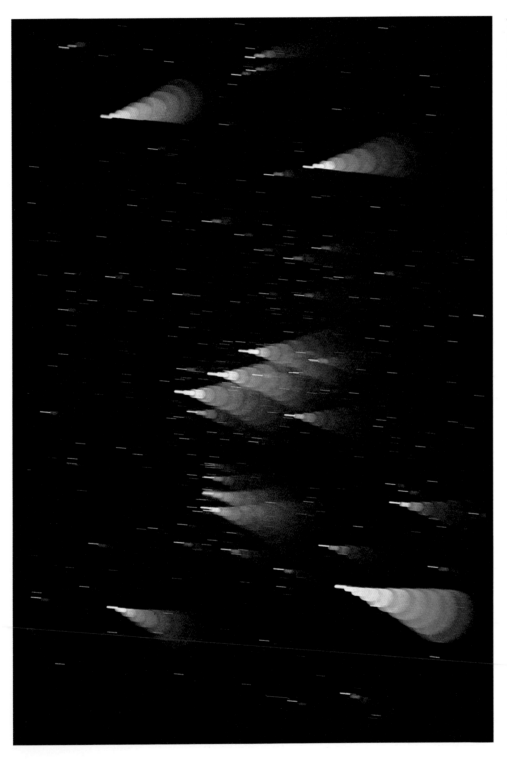

The constellation of Orion is the showpiece of the Milky Way's Orion Arm. Unlike most constellations, the stars in Orion (with the exception of Betelgeuse, top left) are all situated at roughly the same distance – 1600 light years away. This unusual photograph of Orion was made by changing the position of the lens focus nine times over the period of the exposure. The result highlights the colours of Orion's stars, and reveals that most of them are blue – which means that they are young and hot. The exceptions are nearby Betelgeuse (a red giant star close to the end of its life), and the Orion Nebula (below the three stars of the belt), which glows pink in the light of ionized hydrogen gas.

lit up by nearby stars. The presence of the dust means that these stars must be young, but they are not massive and hot enough to excite the gas around them. The survey of reflection nebulae showed many of these young stars strung along the wings of the Dove.

American astronomer William Herbst was intrigued why these young stars – only a third of a million years old – are strung along the edge of the nebula. Other observations showed that the Dove was merely part of a larger shell of gas, and a shell that was expanding. According to Herbst 'it is extremely unlikely that the stars would have been created by chance along a 100-light-year arc of an expanding shell 200 light years in diameter'. Almost certainly, the expanding shell had crashed into a stationary gas cloud, compressing it to the point where stars formed.

But why was the nebula expanding? Herbst and his colleague George Assousa believe the most likely culprit is the explosion of a supernova. A supernova that exploded about half a million years ago, they calculate, would have expanded to the size of the Dove, and still be expanding at the rate we observe. And this would be just the right timescale to trigger the birth of stars that are now 300 000 years old.

Although the case for a supernova has not been proven, Herbst and Assousa point out it is difficult to explain the expansion of the shell in any other way. So the Dove Nebula is providing vital evidence to support the theory that the shock from a supernova can produce a new generation of stars.

From the Dove Nebula towards the Sun we find a confusing series of regions where stars are being born – confusing to us mainly because they happen to lie more or less in front of each other as seen from our position in the Orion Arm. As a result, any small errors in measuring their distances can result in a quite different arrangement in space: the map here shows the best estimates we have at the present time, which suggest three quite distinct regions, Monoceros R2, Monoceros OB1 and the great complex of star-birth in Orion.

The survey of reflection nebulae by Sidney van den Bergh and René Racine led astronomers to the first of these three regions of star-birth. In Monoceros, the unicorn, they found a dark cloud where at least 27 stars were just beginning to peek out and light up the surrounding dust. Later observations at wavelengths that penetrate dust have shown that the dark,

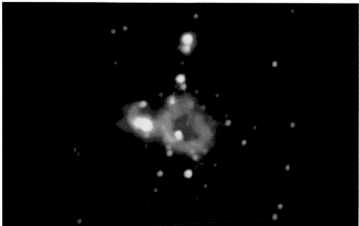

Totally obscured by dust, the star-forming region Monoceros R2 contains one star that is 10 000 times brighter than the Sun. Infrared observations – like the image here – reveal this star, whose radiation makes Monoceros R2 one of the brightest infrared sources in the sky. The star is surrounded by a region of dense dust (shown here), which will soon disperse under pressure from the star's ultraviolet radiation and stellar winds. The complex includes many other young stars that will be revealed as the dust boils away. Dr. Ian Gatley, National Optical Astronomy Observatories/SPL.

dense depths of Monoceros R2 are a hotbed of star-formation – with plenty of stars still to be formed.

The reservoir of gas in Monoceros R2 is shown up by observations of the emission from carbon monoxide, a good indicator of the total amount of dense gas in a cloud. There is enough gas here to form at least 10 000 stars like the Sun. Stretching from this cloud is a strange narrow 'filament' of gas, only 20 light years thick but at least 100 light years long. As we see it in the sky, it stretches out in one direction towards the Orion Nebula and in the other towards Canis Major OB1 and the Dove Nebula. If it really does link these features, all at very different distances from the Sun, then this narrow streamer of gas must be over 1000 light years long!

Infrared astronomers have been able to peer inside Monoceros R2, and have found at least one star that is 10 000 times more luminous than the Sun but totally obscured to optical astronomers by dust. The radiation from this star warms up a surrounding shell of dust, to make Monoceros R2 one of the brightest infrared sources in the sky. One day, its radiation will disperse the dark cloud, and the region will shine as a brilliant nebula.

We can see the future for Monoceros R2 in the next star-formation region along the Orion Arm, where a whole clutch of brilliant young stars has almost succeeded in boiling away the surrounding dark clouds. The association of young stars is known as Monoceros OB1, and it contains some unusual objects: the Christmas Tree Cluster, the Cone Nebula and Hubble's Variable Nebula.

Even a glance through a small telescope at NGC 2264, the main star cluster in Monoceros OB1, reveals how this cluster acquired its nickname. The stars form a symmetrical triangle, like the lights hung on a Christmas tree. At the base of the tree is the brightest star in the cluster, S Monocerotis. This brilliant blue-white star, almost 10 000 times brighter than the Sun, varies slightly in brightness.

The second brightest star in the Christmas Tree Cluster takes the role of the star – or fairy – at the top of the tree. Photographs of this region show something even more interesting. Stretching away from this star is a dark cone of dust, silhouetted against a general background of glowing gas.

The Cone Nebula is a particularly fine example of structures known as 'elephant trunks' or 'cometary globules', that we find in most regions of star formation. The radiation and fast winds from the brilliant young stars forming the Christmas Tree are now sweeping away the remnants of the original cloud. But they have encountered resistance from a particularly dense clump, that now lies at the head of the cone. This clump is protecting the less-dense region that lies downwind, so forming a long column of dark material stretching out from the dense clump, in a direction away from the centre of the star cluster. Long-exposure photographs show several fainter elephant trunks at different locations around the Christmas Tree – all pointing radially outwards.

At the edge of the Monoceros OB1 Association, we come across one of the most peculiar nebulae in the sky. Catalogued as NGC 2261, it forms a small triangle, with at one apex a faint star, R Monocerotis (not to be confused with Monoceros R2!). In the 1860s, astronomers at the Athens Observatory noted that this star varied in brightness – nothing unusual in itself for a very young star. The real shock came in 1916, when the pioneering American astronomer Edwin Hubble compared photographs of the region taken over the previous few years. To his surprise, he found that the nebula itself was changing in brightness and shape.

Hubble's Variable Nebula has fascinated astronomers ever since. Carl Lampland at the Lowell Observatory, for example, studied almost a thousand photographs of the nebula taken over several decades. Hubble had thought that the nebula was changing because lumps of gas were actually moving and varying in brightness, but Lampland found that this could not be true. Sometimes quite large regions would change in just a few days, which meant that the gas in the nebula would have to be travelling faster than the speed of light.

Instead, according to Lampland, the changes in the nebula are a cosmic shadow play. The triangular nebula itself does not change: it is simply a screen that is lit up by the star R Monocerotis. Like many other young stars, R Monocerotis is surrounded by tatters of dust from its birth. This dust explains, first of all, the variable brightness of the star itself: as different clouds of dust move directly between the star and us, they dim the star by an amount that is constantly changing.

The moving dust clouds also throw huge shadows onto the nebula to the side of R Monocerotis. As the dust clouds move, the patterns of dark and light on the nebula move too – and at a much greater speed, just as you can use a slide projector to throw a magnified image of your hands onto a screen to create the face and jaws of a vast dark monster, and each tiny movement of a finger is magnified on the screen.

The star formation regions we've met so far, however, are no match for the great star-factory in our region of the Galaxy: the great star-clouds of Orion. Strictly speaking, Orion – like any other constellation – consists of stars, nebulae and galaxies at very different distances from us, so we should use the name to refer to a region in our sky, rather than a distinct entity in

FACING PAGE Part of the Monoceros OB1 complex, the Cone Nebula is an excellent example of an 'elephant trunk' or 'cometary globule' structure. Winds and radiation from S Monocerotis, at the base of the Christmas Tree, are now sweeping away its natal gas – but the clump below has caused an obstruction. The clump shelters the regions that lie downwind, allowing a column of dark, dense material to survive in the lee of the obstruction. The Cone Nebula is six light years long, and the most prominent member of several other similar structures that point radially back towards the Christmas Tree.

the Galaxy. But Orion is one of those rare cases where most of the objects that are prominent (at any wavelength, from X-ray through visible and infrared to radio) are stars and nebulae that are actually associated in space, as a single complex of star formation, which stretches out along our line of sight from the brilliant star Rigel, at 900 light years, to the great Orion Nebula some 1600 light years from us.

The odd-star-out in the constellation of Orion is Betelgeuse, the bright red star that marks one of the hunter's shoulders. It is comparatively close to the Sun, at only 300 light years, and many books dismiss it as merely a 'foreground' object that happens to lie in the direction of the Orion star-factory. But the star that marks Orion's other shoulder, Bellatrix, is at a very similar distance from us – 350 light years – and yet is exactly the kind of blue-white star that we find in the main body of the Orion star-formation region.

Bellatrix is probably an outlier of the Orion star-factory, left behind as a wave of star-birth swept outwards from the region of the Sun towards the region where the Orion Nebula now lies. So there is every possibility that Betelgeuse too is a true member of the Orion family, but one that has aged slightly more than Bellatrix and has swollen to become a red giant.

If you look towards Orion on an absolutely clear dark night, the character of the constellation changes. A triangular region below and to the right of the Belt becomes a swarm of faint fireflies, just on the limit of vision, pointing like an arrow towards the Sword of bright stars that hangs from Orion's Belt. These 'fireflies' are the young stars of the Orion OB1 Association, a family of stars that have been born in Orion in the past 12 million years – a mere instant in the cosmic scale of time.

Among this family, a few stand out as exceptionally brilliant: shining more brightly than 100 000 Suns, they are among the brightest stars in our part of the Galaxy. This select band includes Rigel and the three stars of Orion's belt, all prominent

naked-eye stars despite their immense distance from us. These are the more massive stars in the Orion OB1 Association. Rigel weighs as much as 30 suns, while the triplet in the Belt may be as heavy as 45 suns. As a result, they have finished the first stage of life, and have expanded to become white-hot giants.

The other stars, at the limit of what we can see with the naked eye, are still the 'main sequence' stars, burning hydrogen at their cores, just as the Sun does. Because they are more massive than the Sun, however, they are correspondingly brighter and hotter, mainly of spectral type B. Turn a telescope on this region, and many more – rather fainter – stars appear. In 1610, the pioneering astronomer Galileo turned his telescope towards Orion, intending to chart the whole constellation as seen through his 'optick tube', but was thwarted when he found that he could see literally hundreds of stars per square degree of sky.

Galileo, oddly enough, did not spot what has become the most famous inhabitant of the constellation: the great Orion Nebula. That honour fell to Nicolas Peirasc, a French astronomer inspired by Galileo. In November 1610, Peirasc found that the central star in Orion's Sword was immersed in 'a small illuminated cloud'. A few years later, the Jesuit astronomer Johann Cysat described the nebula as 'a diffused light like a radiant white cloud'.

But the nebula did not immediately become a household name to astronomers. In 1656, the Dutch astronomer Christiaan Huygens 'discovered' it again. 'There is one phenomenon among the fixed stars worthy of attention, which as far as I know, has hitherto been noticed by no-one', he declaimed. The stars in the centre of Orion's Sword 'shone through a nebula so that the space around them seemed brighter than the rest of the heavens which appeared quite black, the effect being that of an opening in the sky through which a brighter region was visible'. Given what we now know about the Orion Nebula – that it is a bright 'blister' on the front of a vast dark cloud – this was a remarkably prescient comment.

In the shorter term, however, the opinion that most influenced astronomers was that of William Herschel. 'It is an unformed fiery mist, the chaotic material of future suns'. The idea that the material in the Orion Nebula was condensing to form new stars received a boost in the 1880s, when the pioneer of spectroscopy in astronomy, William Huggins, found that

FACING PAGE One of the most classic astrophotographs ever taken, this image of the Orion Nebula by David Malin reveals the incredible complexity of its wispy structure. It was made by combining three plates from the UK Schmidt Telescope, and the details were enhanced by Malin's 'unsharp masking' technique. The pinkness of the main nebula, which comes from glowing hydrogen gas, makes a striking contrast with the blue reflection nebula above.

BOX 1. **Planetary nebulae: death of other suns**

In 1779, a French astronomer called Antoine Darquier was hunting for a comet when he stumbled across 'a very dull nebula but perfectly outlined; it is as large as Jupiter and looks like a fading planet'. A few years later, Sir William Herschel – in his thorough surveys of the sky – came across many similar objects. Following Darquier's description, he called them 'planetary nebulae'. We now know that these small nebulae have nothing to do with planets, but are glowing rings of gas that surround dying stars.

A planetary nebula marks the death of a star that is intermediate in mass, like our Sun. Such a star does not explode as a supernova, and so does not produce an energetic supernova remnant that can be picked out at great distances by its powerful emission of radio waves or X-rays. We can only spot planetary nebulae easily when they are comparatively close at hand, and the best studied specimens are therefore all in the Orion Arm. Fortunately, intermediate-mass stars are ten-a-penny compared with the massive stars that explode as supernovae, so planetary nebulae are much more common than supernova remnants and we have plenty to study right on our galactic doorstep. Astronomers have catalogued about a thousand planetary nebulae; the further reaches of the Galaxy probably contain ten times as many again.

These shells of gas can reach a size of a light year or more, and – because they are comparatively close to us – they seem large in the sky. The nearest, the Helix Nebula (NGC 7293) in Aquarius, is half the apparent size of the Full Moon: you can pick it out with good binoculars on a really dark night.

Although all planetary nebulae are basically expanding bubbles of gas, they are far from looking identical. Over two-thirds are more elongated than 'planetary' in shape, and their diverse appearances have led to a wide range of nicknames. The nebula first spotted by Darquier, in Lyra, is one of the simplest in shape, and is known as the Ring Nebula (catalogued as M57). More like planets are the Saturn Nebula (NGC 7009) and the Ghost of Jupiter (NGC 3242). The names of others have more terrestrial associations: the Owl Nebula (M97), the Eskimo Nebula (NGC 2932) and the Dumbbell Nebula (M27). The bizarrely named 'Blinking Planetary Nebula' (NGC 6826) has a faint star in the centre that seems to blink on and off as you glance straight at it and then away again.

In the centre of every planetary nebula – though not always easy to see – is a very hot compact star, the core of the original star that shed its outer layers to make the nebula itself. The central star can have a temperature as high as 500 000 degrees Celsius, and its intense ultraviolet radiation causes the gas in the nebula to glow. Many planetary nebulae are surrounded by faint halos where there is even more gas than in the nebula we see. A good example is NGC 6543, a small bright planetary nebula in Draco. Here the star in the centre contains about two-thirds of the Sun's mass and the glowing nebula itself only one-tenth as much matter as the Sun, but the almost invisible halo may weigh fully as much as the Sun.

Although planetary nebulae are expanding, their motion is quite slow in astronomical terms, 'only' around 20 kilometres per second, as compared with the thousands of kilometres per second that we find for gases rushing away from a supernova. So a planetary nebula is in no sense an explosion – more a gentle 'puffing away' of gases. After some 50 000 years, this wraith of a star will dissipate into space, leaving only the hot core on view, as a white dwarf star.

One day – in around five billion years from now – our Sun will suffer this fate. Its final few stages will be marked by its own distinctive planetary nebula, undoubtedly christened with a suitable nickname by whatever nearby civilization watches the demise of our star.

FACING PAGE Half the apparent diameter of the Moon in the sky, the Helix Nebula is the closest planetary nebula to the Sun. Its complex double shape is difficult to explain, as are the radial 'jets' on the inner edge of the shell. It is possible that the helical shape represents two ejections from a star in orbit about another star (although there is no evidence that the central star is now double); and that the jets are subsequent, smaller outbursts. This deep photograph reveals glowing material lying far outside the main shell. New measurements reveal that this outer zone contains a great deal of dark material – certainly as much as the matter we can see.

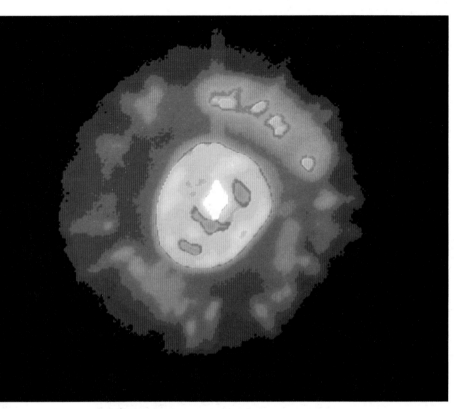

One of the first objects to be imaged by the giant Keck Telescope on Hawaii, the Eskimo Nebula is an unusual planetary nebula lying about 3500 light years away. Surrounding its 'face' is a 'fringe' that resembles the fur on an eskimo's hood. Some astronomers have suggested that the fringe came from an earlier outburst of the central star. As in the case of the Helix Nebula, there is undoubtedly a lot of invisible material surrounding the central bright region.

the nebula is made of gas, rather than being a misty blur of many faint stars like the Andromeda 'Nebula'.

As the twentieth century progressed, it became clear that the hot gases are not condensing. The tight group of young stars in the centre of the Orion Nebula – called the Trapezium, although large telescopes reveal dozens of stars here – are producing strong ultraviolet radiation that is lighting up the nebula, and will eventually disperse the remaining gas. If the bright nebulae were not condensing, then where did stars form? Suspicion fell on dark nebulae that were being revealed

by long-exposure photographs of the sky. Indeed, photographs showed that the whole region around and behind the Orion Nebula was thick with dust.

Orion provided the first testing ground for this theory in the 1960s, when astronomers developed infrared detectors that were sensitive enough to pick up radiation from objects outside the Solar System. Astronomers expected that 'protostars' – stars still in the process of condensing from gas and dust – would emit copious amounts of infrared radiation. The dust in space offers little impediment to infrared wavelengths, so the new instruments should be able to detect star-birth occurring within the dark clouds.

In 1967, two teams of infrared astronomers in the United States picked up infrared radiation coming from within the dark cloud behind the Orion Nebula. Eric Becklin and Gerry Neugebauer found a small source that they thought at first was a protostar. Later research showed that the 'Becklin–Neugebauer object' is in fact a star that has formed, but is still very young and is buried deep in the dark dust cloud. It shines about 10 000 times more brightly than the Sun, but is totally hidden from optical astronomers by the dust.

At about the same time, Douglas Kleinmann and Frank Low found an extended infrared nebula in the same region, but emitting mainly longer infrared wavelengths. The 'Kleinmann–Low nebula' seems to consist of dust heated by young stars in a cluster around the Becklin–Neugebauer object. One of these dust clouds is wrapped around another very young star, which may be as luminous as 100 000 suns.

More sophisticated infrared cameras, developed in the late 1980s, show up hundreds of young stars in the dark cloud of the Orion Nebula. Most of these are comparative heavy-weights, several times the mass of the Sun. When astronomers developed a camera for the UK Infrared Telescope on Hawaii that was sensitive enough to show the birth of stars like the Sun at this distance, they found a surprising dearth of these rather fainter objects. The results back the theory of 'bimodal' star-formation: that heavyweight stars form exclusively in large, very dense clouds like that in Orion, whereas stars like the Sun are born mainly in smaller and less dense clouds.

Radio waves can also penetrate dust clouds, and show us what is going on inside. Most important to astronomers are the specific wavelengths emitted by molecules. In the dense interstellar clouds, individual atoms combine to form mol-

This near-infrared image of the dark cloud behind the Orion Nebula reveals young stars that are totally obscured to optical telescopes by dust. The orange glow at centre is the Kleinmann–Low Nebula, which surrounds a star cluster of very recently formed stars. Among these stars is the Becklin–Neugebauer Object – once thought to be a protostar, but now known to be a fully formed, but very young star. It is 10 000 times brighter than the Sun.

the birth of more stars in the vicinity. It may abort the formation of new stars, by ripping apart clumps of gas that would otherwise have condensed; or it may promote star-formation by compressing lumps of gas that were initially too tenuous to collapse under their own gravity.

The discovery of these fast bipolar flows explained some of the most puzzling features of the regions where stars are born. In the 1950s, George Herbig in the United States and the Mexican astronomer Guillermo Haro discovererd dozens of small bright nebulae in and around the Orion Nebula. At first, astronomers thought that these 'Herbig–Haro objects' were protostars – individual stars in the process of collapsing. But later observations of their spectral lines indicated that the gas in these objects was being heated by powerful shocks, and that each Herbig–Haro object was moving at a high speed – often away from an infrared source that was undoubtedly a protostar or young star. Astronomers began to suspect that a Herbig–Haro 'object' was a region of compressed gas where the bipolar outflow from a young star collided with the surrounding stationary gas.

This theory was confirmed in a spectacular way in 1988 when European astronomers, observing in Spain and Chile, obtained detailed views of several Herbig–Haro objects near the Orion Nebula. HH 111 turns out to be the shock front formed by gases emitted by a hidden star 25 times more luminous than the Sun: the new images revealed the outflowing gases as a 'jet' two light years long, following the line of a bipolar outflow detected at radio wavelengths. Other observations show a jet pointing towards HH 34, from a star of about half the Sun's luminosity. In these images HH 34 is clearly V-shaped, like the bow-shock in front of a boat, and there is a much fainter bow-shock on the other side of the star, as we would expect for a bipolar outflow from the star.

The power of Orion's young massive stars, right on our doorstep in galactic terms, makes this region a striking sight in our skies – both visually and at other wavelengths. As a result, it figures prominently in both professional research papers and popular books as *the* region of star formation. But we should not overestimate its importance in the Galaxy as a whole. The Orion Molecular Cloud is a very average specimen: some of the bigger molecular clouds contain ten times as much matter – over a million solar masses. And there are much bigger and brighter nebulae than Orion, such as the Carina

ecules, and the dark cloud behind the Orion Nebula is now generally known as the Orion Molecular Cloud.

In 1976, a group of American astronomers began to study in detail the emission from carbon monoxide in the Orion Molecular Cloud. They were amazed to find that, in the region of the Kleinmann–Low infrared cluster, the gas was moving at speeds of up to 50 kilometres per second – ten times faster than elsewhere in the cloud. This was the first hint of what astronomers now recognize as a byproduct of the birth of massive stars: a powerful wind of hot but tenuous gas that blows outwards from the energetic young star, in two opposite directions. This 'bipolar outflow' must have a major effect on

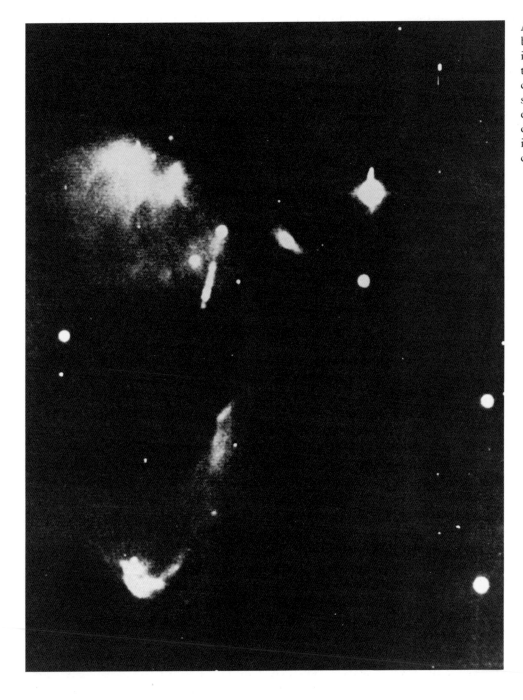

Astronomers originally thought that the 'Herbig–Haro' object HH 34 (bottom of picture) in Orion was a protostar. This photograph from the European Southern Observatory, however, clearly shows that HH 34 is a bow-shaped shock. A young star (near top of the picture) is ejecting a stream of gas, which appears as a downward-pointing jet. The jet becomes invisible as it carries on – until it hits a dense bank of cloud and creates the glowing Herbig–Haro object.

Nebula and NGC 3603, which appear dim in our sky because of their distance and because they are obscured by dust in the Galaxy. If we were flying into the Galaxy from outside, it would not be Orion that would catch our attention: we would head for NGC 3603 or one of the great nebulae on the far side of the Galaxy, which we on Earth know only by obscure catalogue designations such as W49 or W51.

For astronomers confined to Earth, however, Orion has another unique advantage as well as its proximity. Most regions of star-formation lie in the plane of our Galaxy, so that we see them superimposed on more distant objects in the Milky Way. The Orion region lies clear of the band of the Milky Way in the sky, so astronomers do not have to worry that they are confusing objects in the star-formation region with unrelated stars and nebulae lying behind. (Its odd location arises from the fact that the Orion region lies at the edge of a 'fin' of stars and gas projecting from the plane of the Milky Way, Gould's Belt, which we discuss in the next chapter.) As long as we remember that we are looking end-on at this region of progressing star-formation – so that things are generally foreshortened – then we can be pretty sure that what we map in the direction of Orion really does show the relation between different objects in space.

The Infrared Astronomical Satellite, launched in 1983, showed just what a complex tangle of dust clouds fills the Orion region. The main features, however, are quite simple, and also show up in maps made in the radiation from carbon monoxide molecules. There are two large molecular clouds in Orion, each long and thin and containing enough gas to form 100 000 stars like the Sun. As we measure them in the sky, the clouds are each about 150 light years long, but as we are probably seeing them foreshortened, they could well extend for hundreds of light years.

The Orion Nebula lies near the end of one of these elongated clouds, which continues the line laid down by the visible stars of the Orion OB1 Association of stars. This is not mere coincidence. In 1964, the Dutch astronomer Adriaan Blaauw calculated the ages of the stars in various parts of the Orion OB1 Association. At the end furthest from the Orion Nebula, Blaauw found stars around 12 million years old; the stars around the Belt came out at eight million years; those around the Sword at six million years; and the stars intimately involved with the Orion Nebula itself at no more than two million years.

Blaauw concluded that star-birth has occurred in a sequence. It started 12 million years ago and is still progressing eastwards through the constellation (and away from the Sun, if we view the process in three dimensions), with the youngest stars still hidden in the molecular cloud just behind the Orion Nebula. The long molecular cloud stretching away from the Orion Nebula represents a reservoir of material for new stars that will form as the progression of star-birth carries on for the next few million years.

The second molecular cloud in Orion lies near the Belt. In photographs the sharp edge of this cloud shows up clearly, as a boundary between the dark dusty cloud and a luminous sprinkling of distant stars. To one side is sigma Orionis. This appears as a single star to the eye or in wide-angle photographs, but the telescope shows it is a tight cluster of at least five young and brilliant stars, rather like the more famous Trapezium cluster that lights up the Orion Nebula.

Like the Trapezium stars, sigma Orionis is heating and illuminating the gas along the rim of its neighbouring dark cloud. Whereas we see the Trapezium in front of the glowing nebula, here we are observing a view from 90 degrees around, effectively in profile. The nebula appears as a red glowing curtain extending out from the dark cloud, and providing the bright background for a famous silhouette: the Horsehead Nebula.

The Horsehead is a particularly dense clump that projects from the edge of the molecular cloud. As the radiation from sigma Orionis has whittled away the fringes of the large molecular cloud, this clump has been left standing proud. In astronomers' terminology – though mixing the biological metaphors – the Horsehead Nebula is an elephant-trunk structure, similar to the Cone Nebula in Monoceros.

In 1989, a team of American astronomers found the reason why the molecular cloud near the Horsehead Nebula has such an abrupt edge. They mapped the distribution and motion of hydrogen in Orion, by investigating its spectral line at a wavelength of 21 centimetres. Most of the hydrogen surrounded the denser molecular clouds, as the astronomers had expected. But there was also a shell of hydrogen gas lying between the Horsehead and the Orion Nebula. It is growing in size at a rate of seven kilometres per second, and one edge is compressing the dense cloud adjacent to the Horsehead.

This expanding shell is some 50 light years across, making

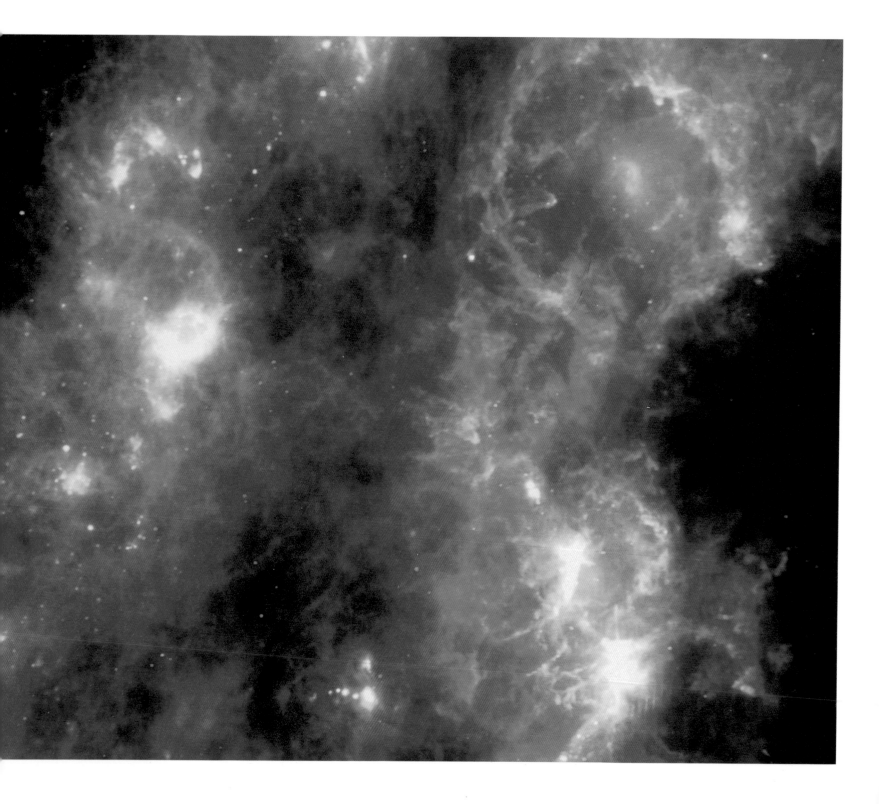

it one of the smallest of many loops and rings of gas in the Orion region. The most prominent ring of glowing hydrogen was discovered in 1894, by the American astronomer E. E. Barnard when he took one of the earliest photographs of the Orion region. Barnard's Loop is a semicircular arc some 300 light years in diameter, with its centre between the Orion Nebula and the Horsehead.

More recent photographs have shown up more structures in the region, including a large loop that stretches into the neighbouring constellation Eridanus, and is hundreds of light years across. Barnard's Loop may just be a bright edge of this Orion–Eridanus shell of gases. It lies on the side of the shell towards the plane of the Galaxy, where we might expect the gases to be compressed – and therefore brighter – as they run into the denser interstellar gases towards the plane.

Astronomers have debated for years about the origin of Barnard's Loop and the other more recently discovered shells of gas. They are undoubtedly expanding, like huge cosmic soap bubbles, but what has inflated them? Some astronomers have thought that the winds of hot gases from the young stars in Orion may be sufficient to blow up these bubbles, but it is more likely that we need the power of repeated supernova explosions. The outward motion of Barnard's Loop shows that it has been expanding for a few million years, and in that time several of the most massive stars born in Orion must have exploded as supernovae.

We can see similar huge shells in the interstellar medium when we observe Orion at virtually any wavelength. The Infrared Astronomical Satellite, for example, showed up a huge 'hood' of warm gas that surrounds the cluster of young stars in Orion's head. Undoubtedly, the interstellar gas in our Galaxy must be riddled with shells and loops like these, but the Orion region provides an ideal place to study and interpret them in detail.

We can, however, pick out several other shells and loops of

interstellar gas in the Orion Arm (indeed, the Sun lies virtually at the intersecton of three such shells, as we will see in the next chapter). Around in the sky from Orion, in the southern constellation Vela, we come across one of the biggest loops in our part of the Galaxy: the Gum Nebula.

As seen from the Earth, the Gum Nebula is the largest object in the sky (apart from the Milky Way itself) at optical wavelengths, stretching across 36 degrees of Vela and its neighbouring constellations. If it were bright enough to be visible to the naked eye this nebula would be a stupendous sight: glowing sheets of gas billowing up from the horizon almost halfway to the zenith, and bringing to life the meaning of the constellation's name Vela, the sails of the great ship Argo – an ancient constellation so vast that Victorian astronomers dismantled it into Vela, Puppis (the poop) and Carina (the keel).

But this is in fact an extremely faint nebula. No-one knew of its existence until 1953, when Colin Gum, a young Australian astronomer, took wide-angle photographs of the southern sky through a filter that passed only light from ionized hydrogen and nitrogen. Gum's pictures were each 11 degrees across, and several showed large sheets of glowing gas. When he assembled the individual photographs as a mosaic, Gum found these sheets joined up to form the enormous, and previously unsuspected, nebula.

Colin Gum was tragically killed in a skiing accident seven years later, long before astronomers reached any consensus about the nature of the nebula he had found. What has become clear is that the Gum Nebula encloses an association of young stars, Vela OB2, which contains some of the hottest stars in the Orion Arm. This group of stars lies 1500 light years away from the Sun – about the same distance as the Orion Nebula. At this distance, we can calculate that the Gum Nebula is over 800 light years across. The most recent measurements show that this huge shell of gas is expanding at a rate of about 10 kilometres per second.

As with the loops in Orion, the Gum Nebula is probably a huge bubble, inflated by the explosion of a succession of supernovae over the past million years. These could certainly account for the size of the expanding nebula. But the gas in the huge shell should, by now, have cooled down and should not be glowing. There are two ways it could have been reheated. The radiation from a recent supernova may have

Hundreds of light years across, Barnard's Loop – photographed here in the red light it emits – encircles the stars in Orion. Astronomers have recently discovered several similar shells of gas, all of which are expanding (there is one surrounding the stars in Orion's head, at top). Barnard's Loop has been expanding for the past few million years, and is probably powered and produced by young massive stars that explode as supernovae.

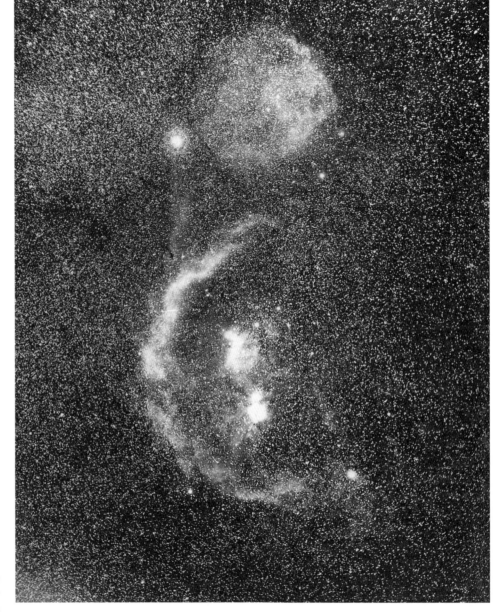

FACING PAGE Three light years from 'nose' to 'mane', the Horsehead Nebula is a denser-than-average lump of gas and dust at the edge of a molecular cloud in Orion. Most of this region has been scoured by powerful winds and radiation pouring off the young star sigma Orionis (off the top of the photograph), so that the area above the Horsehead is largely clear. The winds and radiation also cause the molecular cloud to be compressed – hence its sharp edge.

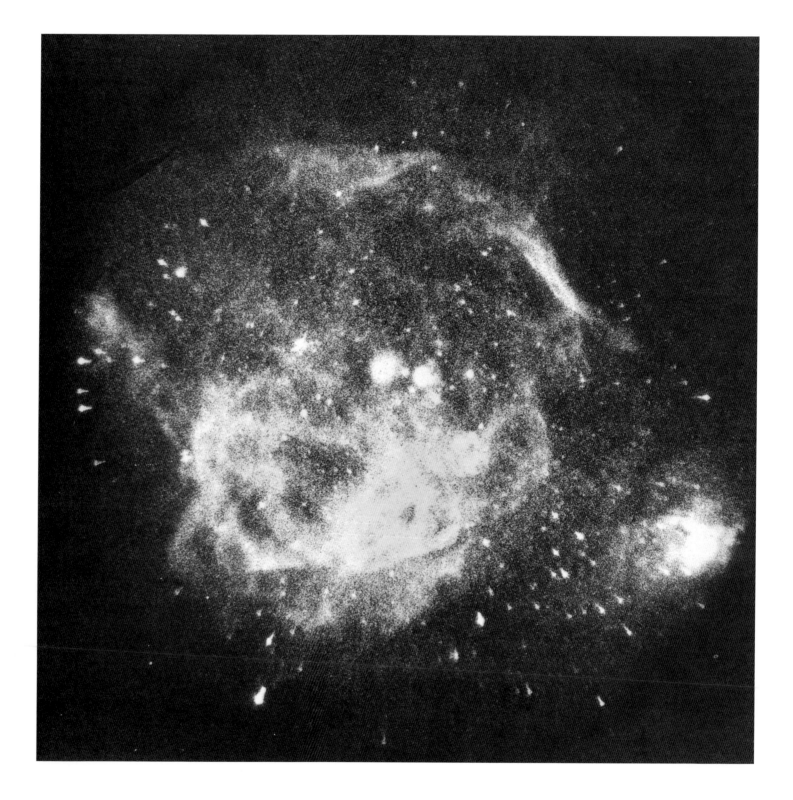

given the gas a brief burst of fresh energy; or the ultraviolet light from extremely hot stars in the Vela OB2 Association may be providing enough energy to make the Gum Nebula shine.

We know that a supernova certainly did explode near the Gum Nebula about 10 000 BC. Astronomers have found a radio source some 5 degrees across, coinciding with a region of very hot X-ray-emitting gas and with a tangled web of fine gas filaments that shows up on long-exposure photographs. These bear all the hallmarks of the gaseous remnant of a supernova, some 12 000 years after the explosion.

In addition, this Vela Supernova Remnant contains a pulsar – a spinning neutron star. Most pulsars emit only radio waves, but this one produces X-rays and light as well. It was only the second light-emitting pulsar to be found, after the pulsar in the Crab Nebula, and when it was discovered in 1977 the Vela Pulsar was the faintest 'star' ever detected. Its emission of light indicates that the Vela Pulsar is very young. This is borne out by its rapid rate of rotation – 11 times per second – and by the fact that its period of rotation is increasing at a substantial rate. From these figures, astronomers can work out that it was created 12 000 years ago: this is our best estimate for the date of the Vela supernova.

There may even be a contemporary record of this supernova, inscribed on stones in the Australian outback. British astronomer Paul Murdin has suggested that 'sunburst' carvings in New South Wales may be the Aborigines' record of this supernova, a brilliant light in the sky that would, for a month, have appeared as bright as the Moon.

Although the supernova in Vela is undisputed, other astronomers do not believe that it was responsible for lighting up the Gum Nebula. They argue that the radiation from the stars in the Vela OB2 Association is quite sufficient. The key to their argument is that this association contains two of the hottest ordinary stars known in the Galaxy: gamma-2 Velorum and zeta Puppis.

To the naked eye, gamma Velorum appears as a single white star near the boundary of Vela with Puppis and Carina: a telescope splits it into two, with some other stars nearby to make 'a most striking field', according to the Australian astronomer E. J. Hartung. Gamma-2 is the brighter of the pair, and a spectroscopic study shows that it is an extremely close double star. The two stars orbit one another every 78 days, and weigh in at 17 and 32 solar masses. Unusually, the less-massive star here is the brighter of the pair – and it has a spectrum that displays broad bright emission lines of carbon and silicon, rather than the narrow absorption lines that we normally find in hot stars.

These characteristics place the lighter component of gamma-2 Velorum in the class of Wolf–Rayet stars: stars that have lost their outer envelopes of gas to expose an extremely bright and hot core. Indeed, the surface temperature of this star comes out to 32 000 degrees Celsius – six times hotter than the Sun's surface.

But even this temperature is surpassed by a star just over the border into Puppis. Leaving aside white dwarfs and neutron stars (the exposed cores of disrupted stars), zeta Puppis is one of the hottest stars we know. The Dutch astronomer Cornelis de Jager has examined several conflicting estimates of its temperature, and concludes that its surface is at an extraordinary 46 000 degrees Celsius.

Zeta Puppis is not a Wolf–Rayet star; it is an ordinary O-type star, and its surface is hot just because it is burning so fiercely. This star is about 80 times heavier than the Sun, and the nuclear fusion reactions in its core make zeta Puppis shine more brightly than a million suns. The star is so hot that most of this radiation is emitted at ultraviolet wavelengths: if our eyes were sensitive to all its emitted radiation, zeta Puppis would appear brighter than Sirius in our skies.

Although zeta Puppis lies within the Gum Nebula, it is a long way from the centre of the Vela OB2 Association of young stars. According to the Dutch Adriaan Blaauw, such 'runaway stars' are the victims of a suicidal companion. Zeta Puppis was once part of a close double-star system, paired up with a star that was even more massive. The two stars were whirling around one another at tremendous speed, until the heavier companion 'went supernova', about a million years

FACING PAGE The Gum Nebula is a huge shell of glowing hydrogen, 800 light years across and filling 36 degrees of the sky in the direction of Vela: this wide-angle photograph covers a region of sky that would stretch from the horizon half-way up to the zenith! The light from the Gum Nebula is so spread out that this structure only appears on wide-angle shots taken though a filter that isolates the light from hydrogen: it is invisible to the eye or to any normal telescope.

The delicate traceries of the Vela supernova remnant are the remains of a star that exploded 12 000 years ago. At the centre of the wreckage is a pulsar that emits light, in addition to radio waves and X-rays – the second ever to be discovered. When the pulsar was found, in 1977, it was the faintest star then detected. It is possible that the supernova itself was witnessed by Aborigines in New South Wales, who carved 'sunburst' symbols on rocks.

ago. With its companion gone – or reduced to a lightweight black hole or neutron star – zeta Puppis shot off like a stone from a sling.

Blaauw first applied this theory to three stars that are racing away from Orion – at such a rate that they now all appear in different constellations in our skies. Hence the variety in their names: mu Columbae, 53 Arietis and AE Aurigae. These are all massive and intrinsically bright stars, able to provide enough ultraviolet radiation to light up any gas that happens to be around. AE Aurigae is indeed passing through an interstellar gas cloud at the moment, and lighting it up as the beautiful wispy nebula IC 405 – more romantically named the Flaming Star Nebula.

Even more spectacular is the effect of the runaway xi Persei, which has escaped from a star-forming region marked by the Perseus OB2 Association and the nebula NGC 1333. An infrared source in this nebula hit the headlines in 1991 as the best contender yet for a genuine protostar. Xi Persei has rushed away from its birthplace, and is now approaching a rippled sheet of dense interstellar gas and dust. The star's radiation is making the gas glow gently. The characteristic shape of the illuminated region has given it the nickname the California Nebula. This nebula is fully two degrees long in the sky, four times the apparent diameter of the Moon. Although it is familiar from photographs, the nebula is quite faint and difficult to see visually.

Some runaway stars are actually leaving a wake as they plough through interstellar gases. Twice as far away as the California Nebula – in roughly the same direction – is another runaway, alpha Camelopardalis. Images from the Infrared Astronomical Satellite show a bow shock around this star. It stands off about 15 light years from the star, and so appears in our skies larger than the apparent size of the Moon. This supergiant star is losing mass at a high rate: in front of the star, this stellar wind piles up to form a stand-off shock, like the shock in front of a supersonic aircraft.

Alpha Camelopardalis is 'running away' from the association Camelopardalis OB1, which lies at the end of a small branch of young stars that runs parallel to the main part of the Orion Arm. Further along this branch, we find two of the brightest red giants in our part of the Galaxy, rho Cassiopeiae and RW Cephei. Rho Cassiopeiae is 600 times larger than the Sun – so vast that Cornelis de Jager classifies it as a 'hypergiant': one of only half a dozen known in our Galaxy.

Coming back towards the Sun, we find a whole rash of star-formation regions about 2800 light years from us, in the constellation Cepheus. There are three main regions, with rather different characters. Cepheus OB4 is a hotbed of star-birth: it contains a dark cloud that is a strong source of infrared and radio waves, powered by the young stars and protostars hidden within. The radiation from these stars has blown a distinct bubble of glowing gas, the 'Cepheus Loop', which is about a degree across. The second association, Cepheus OB3, contains a couple of star clusters that form an attractive sight when seen in a small telescope and a compact region of star formation called Cepheus A.

A remarkable faint and very large nebula encloses the third star association in the constellation, Cepheus OB2. E. E. Barnard discovered this nebula, IC 1396, at the end of the last century, using a 15-centimetre refractor. The nebula is over two degrees across, and so diffuse that it is difficult to see visually, but IC 1396 is spectacular in photographs, with several elephant trunks silhouetted against the glowing background.

What you can see here with a small telescope – or even the naked eye – is a glorious red star. The colour of mu Cephei is so intense that William Herschel dubbed it 'the Garnet Star'. Mu Cephei lies at one edge of IC 1396, and is a brilliant red giant star, 100 000 times brighter than the Sun. It is the last phase in the life of a star in the Cepheus OB2 Association – a star that has experienced serious 'middle-age spread'. As its radius has increased, the star's temperature has dropped to 3600 degrees Celsius, and it shines most intensely at red and infrared wavelengths. Like many red giants, the Garnet Star changes in brightness and colour as its outer layers expand and contract, in a rather irregular manner.

Lying near to the Garnet Star in the sky, but rather closer to the Sun, is the epitome of the *regularly* pulsating giant stars, delta Cephei. The young English astronomer John Goodricke first noticed that this star was variable, in 1784. Despite his handicap of being deaf-mute, Goodricke had already discovered the variability of the light from Algol (see next chapter) and beta Lyrae. His last paper was entitled 'A Series of Observations on, and a Discovery of, the period of the Variation of the light of the Star marked δ by Bayer, near the head of Cepheus'. He found that this star varies regularly over a period of 5 days 8 hours and 37½ minutes (later corrected to

Four times the apparent size of the Full Moon, the faint nebula IC 1396 was discovered by E. E. Barnard at the end of the nineteenth century. It surrounds the association Cepheus OB2, and photographs like this reveal that it is threaded through with elephant-trunk structures. The bright star at the edge of the nebula, also part of the association, is mu Cephei – a star so red that William Herschel nicknamed it 'the Garnet Star'. It is a very luminous red giant star 100 000 times brighter than the Sun, and has cooled as it expanded. Its surface temperature is only 3600 degrees Celsius, which makes it glow garnet-red.

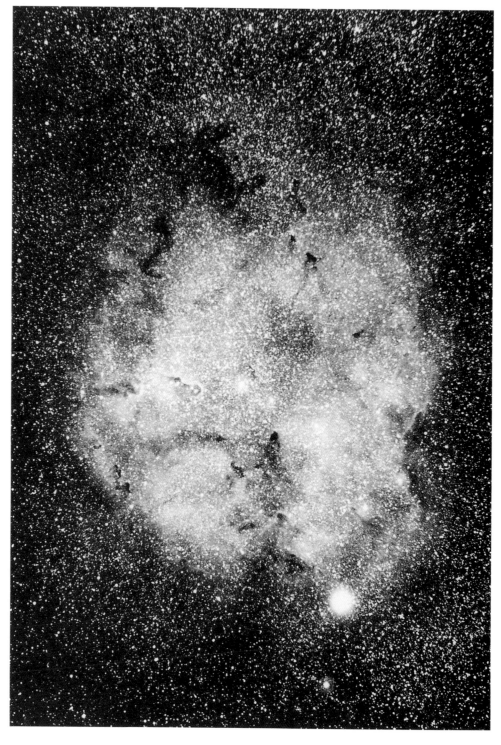

FACING PAGE One of three major star-forming regions in Cepheus, the Cepheus Loop is a huge bubble of glowing gas that surrounds the association Cepheus OB4. In this infrared view from the IRAS satellite, the brilliant young stars in Cepheus OB4 light up their natal cloud. Although it lies 2800 light years away, the Cepheus Loop appears about a degree across in the sky.

45 minutes, which is only two minutes different from modern measurements). While observing delta Cephei in April 1786, however, Goodricke contracted pneumonia and died shortly afterwards, at the age of 21.

Astronomers subsequently found several dozen similar stars – known generically as 'Cepheids' – in our Galaxy. When studying Cepheids in the Small Magellanic Cloud in 1908, the American astronomer Henrietta Leavitt found that the period of variation is related to the star's luminosity – thus allowing Cepheids to be used as yardsticks for measuring the size of the Universe.

A Cepheid is not intrinsically different from other stars. A star that we spot as a Cepheid is just a giant star going through a short phase of its evolution when it tends to pulsate. All stars that are several times heavier than the Sun will become Cepheids for a while in their lifetimes.

In our tour of the Orion Arm, we have skirted the region immediately around the Sun, which includes the young stars of Scorpius, and which we will investigate in the next chapter. We have now reached the region that lies, as seen from Earth, in the opposite direction to Orion and Vela. Here, in the constellation Cygnus, we are looking directly along the Orion Arm as it spirals inwards. As a result, there are plenty of interesting objects in Cygnus, but they all appear projected on top of one another, so it is difficult to make any sense of this region without obtaining accurate distances to all the objects here. Add to this the dense dust clouds that abound here, and we find that this region of our local spiral arm is surprisingly little known.

The Victorian amateur astronomer the Revd T. W. Webb summed up the problem with Cygnus as well as anyone: 'I had at one time projected a survey of the wonders of this region

BOX 2. **Bipolar nebulae: footprints in the sky**

In our corner of the Galaxy, we can spot comparatively faint objects that must be common in the Milky Way as a whole, but are visible only when they are close to us. One of the weirdest of these catagories is the 'bipolar nebula'.

The German-born astronomer Rudolph Minkowski was working at the Mount Wilson Observatory, California, in 1946 when he came across the first example. In the 'neck' of Cygnus, the swan, he found a pair of small faint nebulae, one almost round and the other more elongated – appearing just like the heel and sole of a footprint. Minkowski and his colleagues were perplexed – and the answer only became clear three or four decades later, after astronomers had came across several other bipolar nebulae. Each of these strangely shaped nebulae was so different, however, that it took a long time to make the connection.

In the early 1970s, the US Air Force launched rocket-borne telescopes to survey the sky for natural sources of infrared emission. (This was an important first step in employing infrared detectors to spot incoming ballistic missiles!) One unexpected infrared source lay in the constellation Monoceros. The first photographs of this object showed that it was extremely red in colour and rectangular in shape. Better images of the

Red Rectangle have revealed that it is actually two cones joined at the apex, rather like the central part of an hour-glass.

Another infrared source from the USAF survey turned out to be an egg-shaped nebula in Cygnus. This gastronomic theme was continued with the discovery, on photographs taken from the Cerro Tololo Inter-American Observatory in Chile, of the 'Big Mac' Nebula – two flattened gas clouds slapped together like the buns of a hamburger. Then astronomers at the neighbouring observatory on La Silla found the Calabash Nebula in Puppis, shaped like a drinking gourd.

Despite their diversity of shape, all these nebulae have several things in common. Each consists of two halves (hence the name 'bipolar'), and detailed observations show a faint star in between. And they are all strong sources of infrared radiation, implying that they contain a lot of dust. So astronomers have built up the following picture. The star in the centre is intrinsically a bright and powerful star, but it is largely hidden from our view by a surrounding thick disc of dust that we view edge-on. The star is ejecting gases. Because of the thick dust around the equator, the gas can only escape in the direction of the two poles, and so forms a double nebula.

These unusually active and dusty stars must be either very

young – and still forming from dust and gas – or else very old, and beginning to eject matter into space as they reach their death throes. In the 1980s, the arguments raged between these two exactly opposite interpretations. The situation now seems a lot clearer. Although we do indeed find bipolar nebulae in the large gas clouds where stars are being born, like Orion and the Lagoon Nebula, most astronomers now believe that the isolated bipolar nebulae – like Minkowski's Footprint and the Big Mac – mark the death throes of elderly stars. They are red giants that have become unstable, and are beginning to shed gas from their equators. The heavier elements, such as carbon, condense into grains of solid dust which form the thick disc around the star.

Eventually, the ejected gas and dust from the bipolar nebula will form into a planetary nebula (see Box 1). At first glance, this may seem unlikely, because the most widely photographed planetary nebulae are ring-shaped. These, however, make up only one-tenth of the total; most planetary nebulae are elongated to a greater or lesser extent. The Bug (or Bowtie) Nebula (NGC 6302), for example, looks nothing like a ring: its overall shape resembles a bowtie, but it is ragged at the edges, giving the appearance of a bug's legs and folded wings. It is easy to see how a bipolar nebula could evolve into the Bug.

According to theory, a dying star will appear as a bipolar nebula for a very short period, only a couple of thousand years. This brief appearance explains why they are much rarer than planetary nebulae, which last up to 100 000 years before they fade out: whereas astronomers have catalogued a thousand planetary nebulae, we know of less than a dozen bipolar nebulae.

Minkowski's Footprint, the Red Rectangle and their kin are important snapshots of a fleeting but drastic moment in the life of a star like the Sun. So they are essential to astronomers studying the death of stars. But they also provide a unique laboratory for chemists. Reactions between their gases and the condensing dust grains are forming organic molecules that we do not find elsewhere in space, and which are not common on the Earth.

Bright bands in the infrared spectrum of several bipolar nebulae suggests that the gas contains molecules made up of rings of carbon molecules joined together. The simplest 'ring' molecule is benzene, which consists of six carbon atoms in a hexagon (with six hydrogens attached); if several benzene rings are connected, like a cross-section of a honeycomb, we have a 'polyaromatic hydrocarbon'. Bipolar nebulae contain several polyaromatic hydrocarbons, such as naphthalene, where two benzene rings share a pair of carbon atoms, and coronene, which contains 24 carbon atoms in contiguous rings.

And the very redness which gives the Red Rectangle its name is a clue to more complex chemistry. The optical spectrum shows strong and very unusual emission at red wavelengths, apparently coming from the grains of dust. Years of laboratory research now suggest that the emission comes from grains of carbon with hydrogen atoms bonded onto their surfaces: hydrogenated amorphous carbon. Laboratory experiments show that this material fluoresces strongly, glowly brightly in visible light when illuminated by ultraviolet radiation. The Red Rectangle, therefore, is such a brilliant colour because the material does not just reflect red light. It is actually shining bright red under the influence of the strong ultraviolet radiation from the central star, just as a dancer's teeth fluoresce in the ultraviolet lights at a disco.

with a sweeping power; but want of leisure, an unsuitable mounting and the astonishing profusion of magnificence combined to render [the] task hopeless for me'.

Our first port of call in Cygnus must be Deneb – a beacon 70 000 times brighter than the Sun, and one of the most luminous white-hot (A-type) stars that we know. Deneb is so brilliant that, despite its distance of 1800 light years, it still ranks as a first magnitude star in our skies, outranking literally millions of stars that lie closer to the Earth. The name Deneb means 'tail' in Arabic, and it is quite easy to see how this star forms the tail of Cygnus, the swan. Its body stretches along the Milky Way to a head in the star Albireo, while wings reach out to either side.

Close to Deneb in the sky, and also nearby in space, is a shining gas cloud whose name needs no interpretation: the North America Nebula. You can see this extended patch of

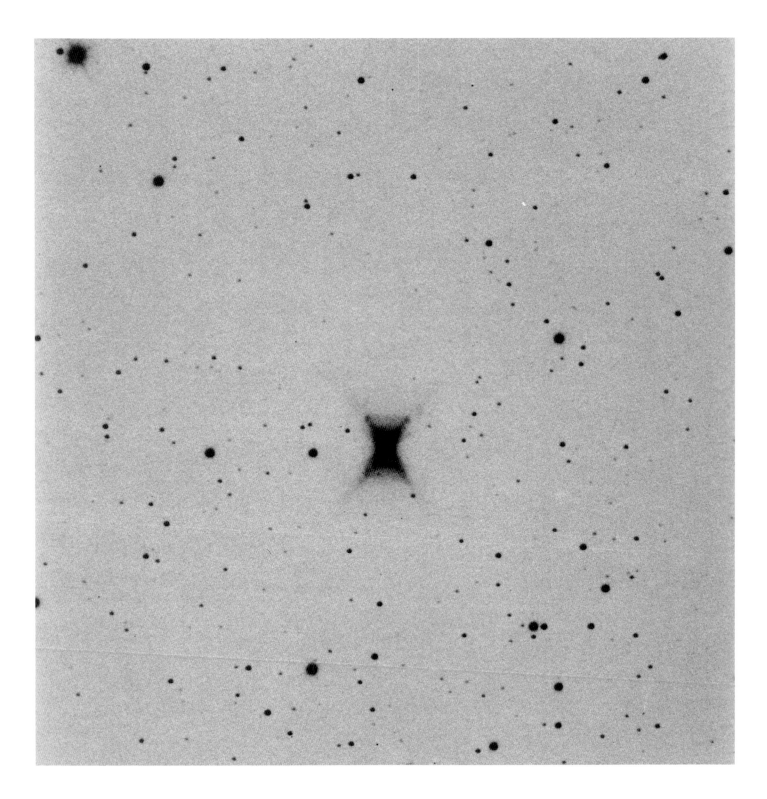

FACING PAGE Almost artificial in appearance, the Red Rectangle is a source of intense infrared radiation. This radiation comes from a thick shell of dust around the equator of an old star, which is ejecting violent winds of gas into space. Because of the obstruction caused by the dust, the gas can only escape at the poles – hence the 'twin cone' appearance of the Red Rectangle. Bipolar nebulae like the Red Rectangle are unstable red giants ejecting matter near the ends of their lives, and will eventually become planetary nebulae. The intense red colour of the Red Rectangle is thought to come from grains of carbon with hydrogen bonded onto their surfaces, which fluoresce when illuminated by the powerful ultraviolet radiation from the central star.

The Bug, Butterfly or Bowtie Nebula (depending on your preference) is a planetary nebula formed from a bipolar outflow. Like the Red Rectangle, it too is very red. Unlike most planetary nebulae, however, the outflow that created The Bug was quite violent. Whereas the speed of most outflows is around a few tens of kilometres per second, the velocities in The Bug are as high as 400 kilometres per second. Additionally, the gas near the centre of the nebula has been highly ionized, indicating much higher energies than are usually encountered in planetary nebulae.

emission – three degrees in length – with the unaided eye on a really dark night. William Herschel first discovered the nebula in 1786, but its true shape only became apparent when Max Wolf photographed the region in 1890. Much fainter is its neighbour, the Pelican Nebula. Parts of it were discovered by the Victorian amateur astronomer T. Espin, but its characteristic shape only appears when the region is photographed.

Radio astronomers have found that the North America and the Pelican are in fact parts of the same glowing nebula, a huge cloud 100 light years across (six times bigger than the Orion Nebula). There is strong radio emission not only from the visible regions, but also from the 'dark space' between the two nebulae, indicating that this is not in fact empty space, but clouds of dark dust (which radio waves can penetrate) hanging in front of the glowing nebulae. In optical photographs, these sheets of dust appear in silhouette as the 'Atlantic Coast' and 'Caribbean Sea' of the North America Nebula. Towards the far edge of the dust sheets, some light can struggle through, in patches that make up the shape of the Pelican.

Travelling further along the spiral arm, we come upon a very different kind of nebula, some 2500 light years from the Sun. This nebula, too, is large – covering over two degrees of sky – but here the glowing gas is confined to thin tendrils around the outside. William Herschel first noticed several of these segments, but it required wide-angle photographs to show the overall shape. Detailed photographs show the gas lies in thin gauzy sheets, leading to names such as the Cirrus Nebula and the Veil Nebula. But these romantic ideas have now been largely replaced by astrophysicists who want to explain the overall structure of the nebula, and it now goes under the prosaic title 'the Cygnus Loop'.

As well as producing light, the Cygnus Loop is a strong source of radio waves, X-rays and infrared emission. It is also expanding outwards, at a speed of 100 kilometres per second. The only reasonable explanation is that the Cygnus Loop is the remnant of a supernova, which exploded about 20 000 years ago. While younger remnants, such as Tycho's supernova remnant in the Perseus Arm, are so hot that they emit very little visible light, the gases in the Cygnus Loop have cooled down to about 10 000 degrees Celsius – and so shine at optical wavelengths.

Slightly further from the Sun, we find the remains of a much less powerful stellar explosion. Nova Cygni blazed out in the summer skies of 1975, rivalling Deneb in brightness and reaching a total luminosity of about a million Suns. Amateur and professional astronomers alike homed in on this brilliant sky-sight.

Like other novae, this eruption did not mark the death of a star. It was a superficial explosion of gases that had spilt off one member in a close double-star system, and accumulated on the surface of the companion, a white dwarf. Eventually, the accumulation of hydrogen-rich gas exploded. From being intrinsically fainter than the Sun, the double star system became for a few days the brightest object in the Orion Arm.

Near the edge of the region we are surveying in this chapter, we find a star that is actively jettisoning much of its matter without exploding. The gases from the star – known only by its catalogue number HD 192163 – have formed a distinctive egg-shaped nebula, NGC 6888, which is visible in a small telescope even though it lies 4000 light years away. The star HD 192163 is a Wolf–Rayet star, like gamma-2 Velorum, and shows abnormally strong lines of nitrogen in its spectrum. Astronomers have hypothesized that the surfaces of these stars have an unusual composition because they have lost their outer envelopes of hydrogen: in the case of HD 192163 we can actually see the lost envelope, in the shape of the nebula NGC 6888. The star has probably been losing matter for about 20 000 years, and the nebula now contains about as much matter as the Sun.

Next to NGC 6888, but not associated with it, wide-angle photographs of Cygnus show a whole complex of nebulosity covering many square degrees of sky. In the centre lies the star gamma Cygni, which marks the centre of the constellation Cygnus. Optical astronomers have long called this the 'gamma

FACING PAGE The 'tail' of Cygnus the Swan is wreathed in dust and nebulosity. To the left of the bright star Deneb (top centre) is the amazingly lookalike North America Nebula. Right next to it is the fainter Pelican Nebula (which really does resemble a pelican on long-exposure photographs). The sharp boundaries of these nebulae (like the 'Gulf of Mexico') are caused by dark dust, and the whole complex covers an area 100 light years across. On the far right of this photograph is the intricate nebulosity that appears to surround the star gamma Cygni (bottom right) – but these are in fact several very distant nebular complexes that lie along the same line of sight as we observe this part of our local spiral arm 'end-on'.

BOX 3. **Epsilon Aurigae: from 'biggest star' to biggest enigma**

On winter evenings in the northern hemisphere, the bright star Capella is almost overhead. Near it in the sky, but almost fifty times further away, is the star epsilon Aurigae. Although little known outside astronomical circles, this is one of the most intriguing stars in our part of the Galaxy.

In the last century, astronomers found that epsilon Aurigae varies in brightness: every 27 years, its light fades and remains dim for two years before coming back to its normal brightness. The star's spectrum showed the reason for these changes. It revealed that the star is swinging about in an orbit that lasts 27 years, around a star that is much fainter and does not contribute a noticeable amount of light to the spectrum. For two years in each orbit, the fainter star passes in front of the brighter and blocks out its light, so the whole system becomes a lot dimmer. Variable stars are two-a-penny in the Galaxy, and eclipsing binaries are not uncommon, either. But epsilon Aurigae is different.

First of all, we can work out the size of the stars' orbits around one another, and their speeds. From the length of time the dimmer star hides the brighter one, we can easily calculate the size of the 'secondary' star. It comes out as some 3000 million kilometres in diameter – larger than the orbit of Saturn around the Sun, and far bigger than any other star ever measured. From the 1930s to the 1950s, therefore, astronomers gave epsilon Aurigae the first prize for sheer size.

But a deeper analysis of the system, during its 1955–7 eclipse, showed that this theory must be wrong. The spectrum of the bright star shows it is a supergiant, weighing between 15 and 30 times as much as the Sun. From this star's mass, and the details of the orbit, we can calculate that the mass of the dimmer companion is about 10 Suns. Such a star should be quite bright in its own right – yet there are no signs that it is contributing to the total light from the system, even during the eclipse when we would expect to see little or no light from the brighter star and most of the light should be coming from the dimmer star.

Indeed, during the eclipse the spectrum of the system remains exactly the same – indicating that we are not seeing the secondary star at all, but a diminished amount of light from the bright star. Either the secondary star is semi-transparent and we see the brighter star through it, or the secondary is cigar-shaped so that it hides only half the light from the bright star as it gradually creeps across. In our experience, however, stars are neither transparent nor elongated – so what *is* the secondary of epsilon Aurigae?

Most astronomers now favour the idea of an 'elongated' object, but not a star. Instead, the obscuring object is probably a large dark disc of gas and dust that surrounds the secondary star itself. We see the disc at a shallow angle, so the dust in it hides the central star from our view, and the disc itself appears, by projection, as a cigar shape.

This still leaves the question of why the star has a surrounding disc. And there is no agreement yet. Some astronomers believe that the disc of matter is condensing to form a new planetary system around the hidden secondary star, although it is not clear why the secondary star should be at such a youthful stage when the brighter star has become a supergiant star, apparently near the end of its life.

Alternatively, the gases in the disc may have splashed over from the big bright star, to swirl around the secondary star. In other systems where this happens, however, the gases in the disc spiral inwards onto the secondary star and generate a powerful output of X-rays, and we do not pick up X-rays from epsilon Aurigae. We can 'save' this theory, however, if the 'secondary star' actually consists of a close pair of stars. Their orbital energy prevents the gas from dropping down onto either of them.

To fit the observations completely, we need further refinements to either theory. For example, the system brightens up slightly right in the middle of the eclipse, so we probably need to invoke a hole – or a partial clearing – in the centre of the opaque disc. And the way the star fades and brightens tells us that the projected shape is more rectangular than cigar-shaped: we can achieve this either by assuming that the opaque disc is warped or that it has a gap (like the gaps in the rings of Saturn) that mimics the effect of a square-ended disc. Astronomers are looking forward to the eclipse of 2009 to sort out once and for all the most enigmatic star in the Orion Arm.

Cygni nebulosity', but this is a misnomer as the star is totally unrelated and lies much nearer to the Sun. The whole nebula glories in the official name IC 1318. Much of its light comes from gas clouds located about 3700 light years away from us – but IC 1318 also comprises many other nebulae lying at very different distances along our line of sight.

The optical nebulae are dimmed (and confused) by patches of dust lying in front, but radio waves and infrared have no such problems. These wavelengths reveal this region as a brilliant patch of glowing nebulae and young stars. In the 1950s, radio astronomers suggested that this region of sky, which they called Cygnus X, was not a single radio source, but our view of the local spiral arm seen end-on.

We can now begin to tie up our picture of the Cygnus region. Optical astronomy shows us at least half a dozen associations of young stars, stretching out almost one behind the other to a distance of 7500 light years. A comparison of the light and radio emission from the nebulae show that they range in distance from 3000 light years to at least 12 000 light years. Some X-ray astronomers have suggested that there is a 'superbubble' of hot gas about 7000 light years from us, but the X-ray emission is just as likely to come from the general hot gas in the spiral arm, intensified by our viewpoint looking down the length of it.

The other component of the arm is a string of dark clouds of dust, which are thick with molecules. Optical astronomers see these as black outlines against distant stars: radio astronomers tuned to the radiation from carbon monoxide molecules detect the clouds' silver linings. More important, the radio observations allow astronomers to measure the distances to the clouds, and to identify clouds lying behind one another, because they generally move at different speeds.

In the early 1980s, Thomas Dame and Patrick Thaddeus used a small radio telescope sited on an office block in New York city to probe the great clouds of Cygnus. They found that the massed clouds of Cygnus X range in distance from 1500 to 9000 light years and contain as much mass as five million suns. Coming closer to home, their observations resolved individual clouds, with masses around 100 000 suns.

When we plot out the positions of these dark clouds, we find that they are generally strung along a line that is parallel to the OB associations, but slightly closer to the Galaxy's centre. This is similar to the situation that we find in other galaxies, where the inner edge of a spiral arm is often traced by a line of dark molecular clouds.

These clouds are not something that you need complex equipment to detect. On a dark night, when this region of the Milky Way is high in the sky, take a look at the constellations Cygnus and Aquila. The bright band of the Milky Way seems to split here, into two streams separated by a dark gap. In fact, the 'Great Rift' in Cygnus consists of the dark clouds blocking the light from stars behind.

Look more carefully, and you'll notice that the dark rift is narrow where it starts in the middle of Cygnus. Down towards Aquila it becomes broader, and the rift sweeps northwards out of the Milky Way as a wide swathe in Serpens and Ophiuchus. This is the effect of perspective. Our eyes are following a band of dust as it comes from the distance in Cygnus, to pass close to the Sun in Aquila and Ophiuchus (a story taken up in the next chapter).

A twin to the New York radio telescope, but sited in Chile, has followed this band of dark clouds into the southern constellation of Vela. It has shown at least four clouds with masses of half a million suns each, one behind the other, and ranging in distance from 3000 to 8000 light years. To the naked eye, these clouds are not as obvious as the Cygnus dark rift. The background glow of starlight is fainter, so the cloud silhouettes are less obvious, and the region is spangled with the nearby bright stars of the Vela OB2 Association.

Despite the common name for our region of the Galaxy, then, the bright lights of Orion are actually something of a distraction. We should turn to the Great Rift of Cygnus for a literal sign in the heavens. As it snakes towards us through space, this long dark band of dust gives us intuitive proof that we reside within the spiral arm of a galaxy.

CHAPTER 6

Our local neighbourhood:
a typical corner of the Galaxy

AQUILA RIFT

Beta Arae

Dumbbell Nebula

Upper-Scorpius OB Association

Pi Scorpii

Sigma Scorpii

Lupus Dark Cloud

Rho Ophiuchi Dark Cloud

Rho Ophiuchi

Upper Centaurus-Lupus OB Association

Sadalsuud

Zeta Ophiuchi

S27

Dschubba

Graffias

G317 Dark Cloud

Rasalgethi

Chameleon I Dark Cloud

Beta Sagittae

Kappa Centauri

Sadalmelik

Alpha Sagittae

Hadar

Coalsack

Pi Sagittarii

Antares

Eta Centauri

Helix Nebula

Peacock

Shaula

Mimosa

Lower Centaurus-Crux OB Association

Pi Herculis

Phi Sagittarii

Acrux

IC 2602

Nu Coronae Borealis

Zeta Coronae

Theta Carinae

Aspidiske

Enif

Albireo

Borealis

Nunki

R Hydrae

Gamma Cygni

Zeta Cygni

Alpha Arae

Gamma Lupi

Beta Crucis

Sulaphat

Spica

Rastaban

R Aquarii

Avior

Nu Carinae

Kappa Velorum

RR Lyrae

Omicron Velorum

SS Cygni

Scheat

The Sun

Achernar

LOOP I

IC 2391

Alpha Reticuli

Lambda Velorum

Chi Carinae

Alkaid

Sirius Supercluster

Hyades

LOOP III

Algenib

Elnath

Furud

Mu Leporis

Beta Cephei

Eta Aurigae

LOCAL BUBBLE

Nihal

Epsilon Leonis

Iota Aurigae

Adhara

Delta Persei

Lambda Tauri

Betelgeuse

NGC 2451

Polaris

Bellatrix

Epsilon Cassiopeiae

Pleiades

Praesepe

Xi Puppis

S184

Gamma Cassiopeiae

Perseus OB3

T Tauri

Taurus

HL Tauri

CW Leonis

Mirfak

Dark Cloud

Zeta Tauri

Mirzam

Owl Nebula

Zeta Aurigae

Mira

Beta Monocerotis

Eta Persei

LOOP II

Medusa Nebula

Arneb

LINDBLAD RING

Mebsuta

Rigel

Eta Orionis

A

50 Light Years

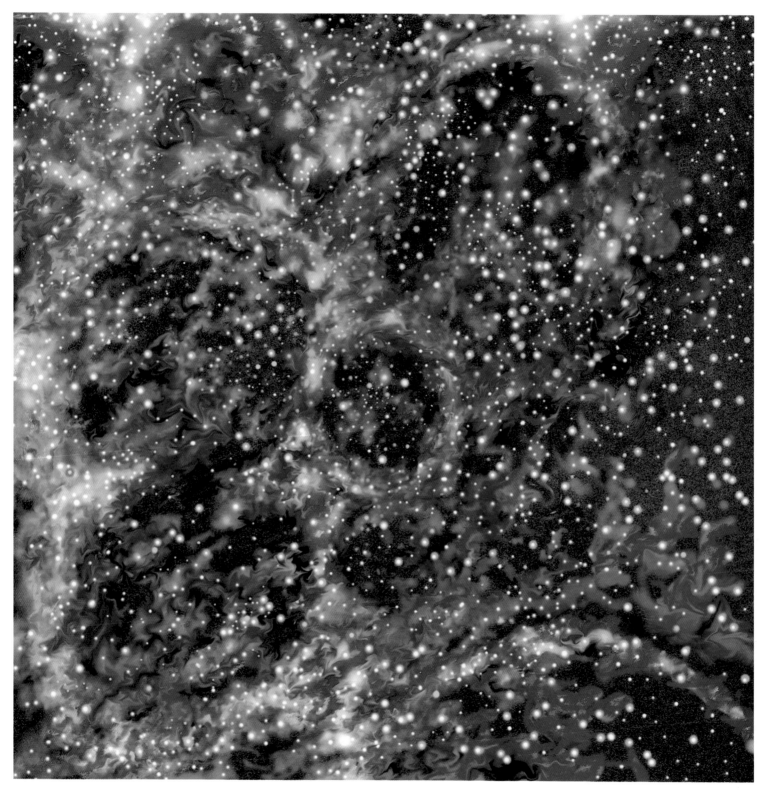

B

KEY TO MAP A ON P. 162

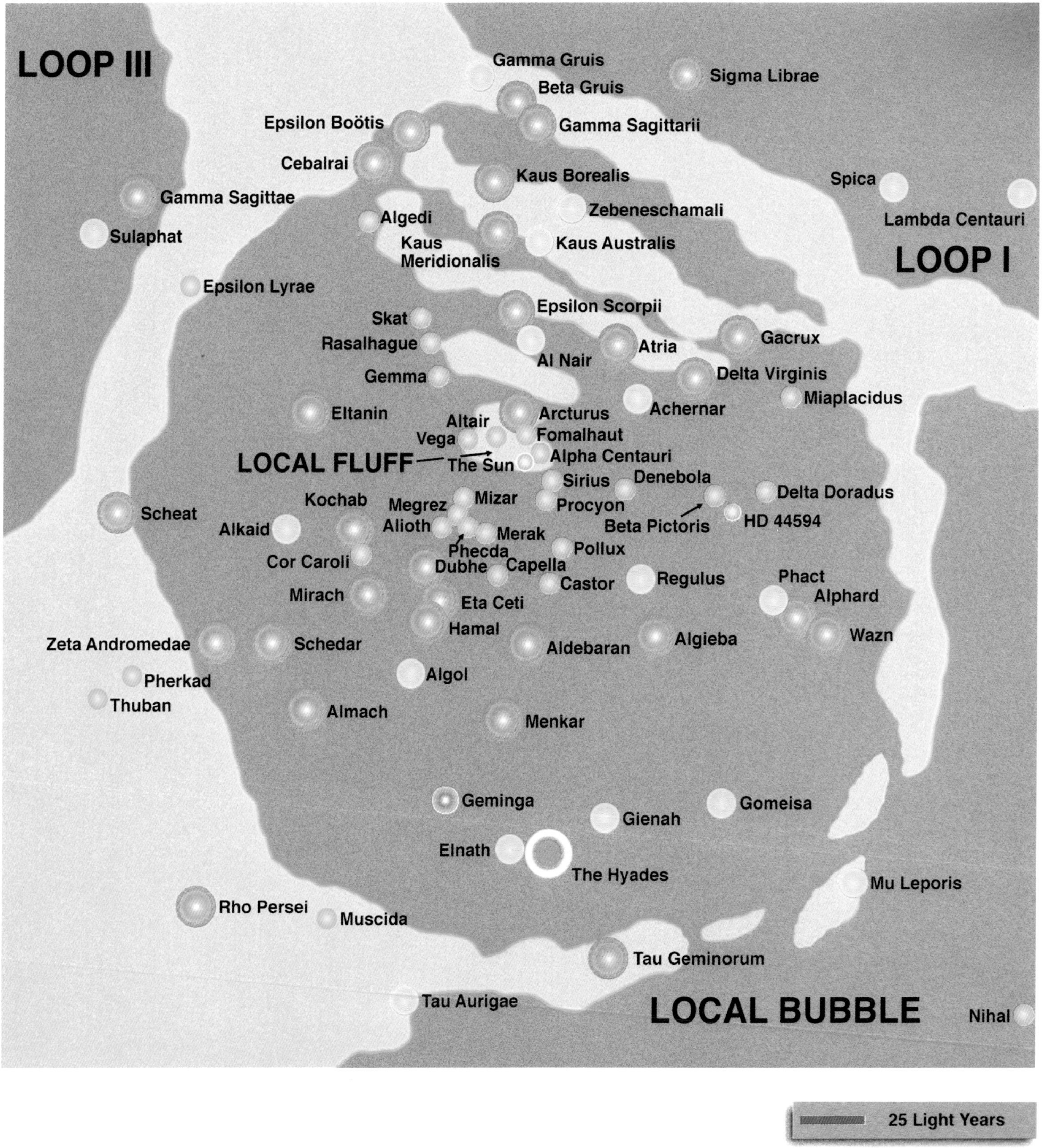

LOOP III

Gamma Gruis

Sigma Librae

Beta Gruis

Epsilon Boötis

Gamma Sagittarii

Cebalrai

Kaus Borealis

Spica

Gamma Sagittae

Zebeneschamali

Lambda Centauri

Sulaphat

Algedi

Kaus Australis

LOOP I

Kaus
Meridionalis

Epsilon Lyrae

Epsilon Scorpii

Skat

Atria

Gacrux

Rasalhague

Al Nair

Delta Virginis

Gemma

Miaplacidus

Eltanin

Altair

Arcturus

Achernar

Vega

Fomalhaut

LOCAL FLUFF — The Sun

Alpha Centauri

Sirius

Denebola

Delta Doradus

Scheat

Kochab

Mizar

Procyon

Alkaid

Megrez

Merak

Beta Pictoris

HD 44594

Alioth

Pollux

Cor Caroli

Phecda

Dubhe

Capella

Regulus

Phact

Mirach

Castor

Alphard

Eta Ceti

Algieba

Wazn

Zeta Andromedae

Hamal

Schedar

Aldebaran

Pherkad

Algol

Thuban

Almach

Menkar

Geminga

Gomeisa

Gienah

Elnath

The Hyades

Mu Leporis

Rho Persei

Muscida

Tau Geminorum

Tau Aurigae

LOCAL BUBBLE

Nihal

C

25 Light Years

D

KEY TO MAP C OVERLEAF

KEY TO MAP A

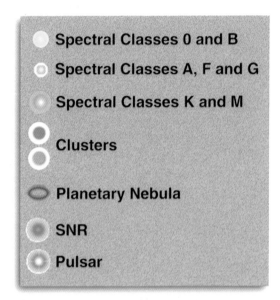

- Spectral Classes 0 and B
- Spectral Classes A, F and G
- Spectral Classes K and M
- Clusters
- Planetary Nebula
- SNR
- Pulsar

KEY TO MAP C

- Spectral Classes O and B
- Spectral Classes A, F and G
- Spectral Classes K and M

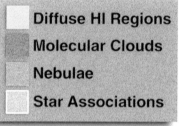

- Diffuse HI Regions
- Molecular Clouds
- Nebulae
- Star Associations

MAP A Almost all the familiar stars lie within a few hundred light years of the Sun, within the tiny portion of the Galaxy covered by this chart. Part of the Local Arm's main dust lane (running across the top right) has condensed into a large OB association that is inflating a giant bubble of interstellar gas, which appears in our sky as the huge radio-emitting Loop I. To the other side of the Sun is the nearest star-nursery, the Taurus Dark Cloud.

These interstellar structures have been deduced from the size and position of the radio-emitting loops, absorption by dust clouds and the distance to stars associated with star-formation regions: they combine to give a coherent picture of young stars blowing bubbles that in turn compress dust clouds to form yet more young stars.

Key to less-familiar star-names:

Acrux	Alpha Crucis
Adhara	Epsilon Canis Majoris
Albireo	Beta Cygni
Arneb	Alpha Leporis
Aspidiske	Iota Carinae
Avior	Epsilon Carinae
Dschubba	Delta Scorpii
Elnath	Beta Tauri
Enif	Epsilon Pegasi
Furud	Zeta Canis Majoris
Graffias	Beta Scorpii
Hadar	Beta Centauri
Mebsuta	Epsilon Geminorum
Mimosa	Beta Crucis
Mirfak	Alpha Persei
Mirzam	Beta Canis Majoris
Nihal	Beta Leporis
Nunki	Sigma Sagittarii
Peacock	Alpha Pavonis
Rasalgethi	Alpha Herculis
Rastaban	Beta Draconis
Sadalmelik	Alpha Aquarii
Sadalsuud	Beta Aquarii
Scheat	Beta Pegasi
Shaula	Lambda Scorpii
Sulaphat	Gamma Lyrae

MAP B Deep inside the Local Arm, we can discern a 'butterfly wing' of bright young blue stars, to either side of the Sun. Their powerful winds are blowing bubbles in the interstellar gas. Scattered among the blue stars are red giants and pink planetary nebulae, marking the final stages of star-life.

MAP C The Sun and many of the most familiar stars lie within the Local Bubble, a tenuous region of space that was probably cleared by the explosion of a supernova. The region mapped here is about the same extent as the thickness of the Galaxy's disc, so these stars actually fill a three-dimensional volume as thick as the width of the page: as a result, some stars appear here much closer together than they are in reality.

The rapidly expanding Loop I is bearing down on the Local Bubble, pushing streamers of gas ahead – including the Local Fluff that is now sweeping past the Sun. The faint star HD 44594 is a 'solar twin', equal to our Sun in mass, brightness and age. Though it is so close to the Sun, it is hardly visible to the naked eye: conversely, even from this comparatively nearby star, the Sun would be on the point of invisibility. The Sun and its twin are marked here only for reference: in fact, they are far fainter than any of the other stars shown.

Key to less familiar star-names:

Algedi	Alpha-1 Capricorni
Algieba	Gamma Leonis
Almach	Gamma Andromedae
Al Nair	Alpha Gruis
Alphard	Alpha Hydrae
Atria	Alpha Trianguli
Cebalrai	Beta Ophiuchi
Cor Caroli	Alpha Canum Venaticorum
Denebola	Beta Leonis
Elnath	Beta Tauri
Eltanin	Gamma Draconis
Gacrux	Gamma Crucis
Gemma	Alpha Coronae Borealis
Gienah	Gamma Corvi
Gomeisa	Beta Canis Minoris
Hamal	Alpha Arietis
Kaus Australis	Epsilon Sagittarii
Kaus Borealis	Lambda Sagittarii
Kaus Meridionalis	Delta Sagittarii
Kochab	Beta Ursae Minoris
Menkar	Alpha Ceti
Miaplacidus	Beta Carinae
Mirach	Beta Andromedae
Nihal	Beta Leporis
Phact	Alpha Columbae
Pherkad	Gamma Ursae Minoris
Rasalhague	Alpha Ophiuchi
Scheat	Beta Pegasi
Schedar	Alpha Cassiopeiae
Skat	Delta Aquarii
Sulaphat	Gamma Lyrae
Thuban	Alpha Draconis

MAP D No blazing supergiants or brilliant nebulae light up our very local neck of the cosmic woods. Our immediate neighbours are fairly ordinary stars, one star cluster (the Hyades), one neutron star (Geminga) and numerous wisps of interstellar gas. This view is typical of most small regions of our Galaxy – and any other spiral galaxy.

Our tour of the Milky Way Galaxy now brings us to a tiny and totally insignificant region. It is only 1500 light years across – no larger in comparison to the Galaxy as a whole than this letter 'O' compared with a dinner plate. This region contains no powerful or exotic beasts that would draw our attention if we were visiting the Galaxy from outside: no brilliant nebulae, no powerful supernova remnants, no voracious black holes.

Despite its lack of any obvious tourist attractions, we are devoting a whole chapter of our guide to this tiny part of the Galaxy. The reason is partly parochial. It is the local neighbourhood of own star, the Sun. Here we find our neighbours in space: familiar objects like the stars Sirius and Antares, the Pleiades and Hyades star clusters and the dark Coal Sack dust cloud.

But astronomers also believe that our local neighbourhood has much to teach us about the Galaxy as a whole. There is no reason to think that we live in some special region of space, so our local region should be fairly typical of any part of the Galaxy. Undistracted by the flamboyant – but rare – objects that call attention to themselves from afar, we can study a cross-section of the comparatively humdrum objects that form the silent majority of the Galaxy's inhabitants – including the structures that gas and dust form in space. In the local neighbourhood, we see these nearby objects in unusually intimate detail. We can also pick out the dimmest of stars, which are invisible beyond a few dozen light years, and hope to find traces of still-fainter objects such as 'brown dwarfs' and families of planets circling other stars.

The stars near the Sun are far from being a family group. They constitute a jumble of stars of all ages, with different backgrounds. Although our snapshot of the Galaxy at the present time shows all these stars in the Local Arm, most of them – including the Sun – are just passing through. Almost 99 per cent of the present inhabitants predate the Local Arm in its present form: only stars less than 30 million years old are associated with the current structure of the arm, as delineated by the concentrations of gas and dust.

The oldest stars in our local neighbourhood belong to the Galaxy's halo, and just happen to be travelling through the galactic disc close to the Sun. These faint stars first draw attention to themselves by their high speeds. They are whizzing past the Sun at velocities up to 200 kilometres per second in the general direction of the constellation Vela. The Swedish astronomer Bertil Lindblad first realized that these 'high velocity stars' are not really streaming past us: they are stars of the halo that are travelling relatively slowly in random directions. The Sun, and the neighbouring stars of the Galaxy's disc, are in fact the stars with the high velocity: we are orbiting the galactic centre at 200 kilometres per second in the direction of Cygnus, and leaving the halo stars behind.

The middle-aged inhabitants of the region include our Sun. During its 4600-million-year lifetime, the Sun has passed in and out of spiral arms dozens of times. It is purely coincidental that we currently find ourselves within an arm – though, as we shall see later, in a rather empty region.

Stars of around one-tenth the Sun's age were also born elsewhere, and have passed through spiral arms several times before their current passage through the Local Arm. Many of these stars are travelling together, in loose swarms several hundred light years across. The members of each swarm – or supercluster – were born together, but are now gradually drifting apart. A supercluster is so large that its stars are at first sight indistinguishable from the many other stars – older and younger – that occupy the same region of the Galaxy's disc, including the members of other, overlapping, superclusters. The only way to sort out the membership is to measure the stars' motions: the stars in a supercluster are all heading in the same direction, at the same speed. So only in the region immediately around the Sun can we begin to pick out the superclusters at all.

In the past century, astronomers have discovered that the Sun is situated within at least three superclusters of stars. The oldest is a swarm of stars that was born along with the Hyades cluster, about 630 million years ago. As well as the Hyades – the stars that make up the head of Taurus, the Bull – this supercluster contains the star cluster Praesepe, in Cancer. All these stars are heading away from us, in the direction marked in our sky by the constellation Orion. The motion of the Hyades – the nearest star cluster to the Sun – provides a method of measuring distances that is independent of the parallax method, and the distance to the Hyades is the first 'rung' in a ladder of distance measurements that takes us out to the furthest galaxies (see Box 3 in Chapter 3).

A younger supercluster surrounds the Sun more evenly, and contains several familiar stars that lie in very different parts of the sky. The discovery came as a surprise to the Danish astronomer Ejnar Hertzsprung, early this century. He found that many apparently unrelated stars – including Sirius, the five central stars of the Plough (Big Dipper) and Gemma (the brightest jewel in the Northern Crown, Corona Borealis) – are all moving in the same direction through space. They are the most prominent of the hundred or so stars making up the Sirius supercluster, which is 490 million years old and is heading past us in the direction of the galactic centre.

The youngest of the superclusters surrounding the Sun is centred on the Pleiades star cluster, some 410 light years away and 70 million years old. This beautiful little group of stars has fascinated people all round the world from earliest times:

the Chinese recorded the Pleiades as early as 2357 BC. It is one of the few astronomical names to crop up in the Bible, as God asks Job: 'canst thou bind the sweet influences of Pleiades?' The Polynesians used the Pleiades as an important navigational aid as they traversed thousands of kilometres of the empty Pacific. To them, these stars were the eyes of Rigi, a worm-god who tried to raise the heavens but broke into pieces under the strain.

The Maya and Aztec people of Central America regarded the Pleiades with special awe, timing their most gruesome human sacrifices to coincide with particular appearances of these stars. The British astronomer Michael Rowan-Robinson has noted that the date of Hallowe'en was originally fixed in

Unlike most constellations, several of the stars in the familiar shape of the Plough are genuinely associated. The middle five stars are all moving in the same direction in space – in common with Sirius, and Gemma, the brightest star in Corona Borealis. These stars are all members of the Sirius supercluster, which is 490 million years old and surrounds the Sun in space. Note the naked-eye double star second from the left-hand end – Mizar and its companion Alcor.

FACING PAGES Most beautiful and most famous of all the star clusters in the sky, the Pleiades contains over 200 stars born some 70 million years ago. The wispy dust in the cluster is not 'natal', however: the Pleiades – although young – are sufficiently mature for their original gas cloud to have dispersed. This image by David Malin reveals that a cloud of interstellar matter is sweeping past the Pleiades: the faint pink glow just below the cluster comes from hot hydrogen gas at the fringes of the cloud.

the same way, and concludes 'it is strange that both European and Meso-American cultures should attach such importance to the night when the Pleiades reaches its highest altitude at midnight'.

The appearance of the Pleiades has been expressed most clearly – and hauntingly – not by an astronomer but by a poet. 'Many a night I saw the Pleiads, rising thro' the mellow shade,/ Glitter like a swarm of fireflies, tangled in a silver braid,' declared Alfred, Lord Tennyson, in Locksley Hall. In more prosaic terms, the cluster appears to the naked eye as a small group of bluish stars. Although the cluster has been known since ancient Greek times as the 'Seven Sisters', most people see any number but seven! Moderate eyesight shows six stars, people with better vision can see nine or ten, while the keen-eyed Victorian astronomer William Rutter Dawes claimed to see thirteen. The British astronomer Patrick Moore once asked the viewers of his television programme *The Sky at Night* how many Pleiads they could see: the answers ranged fairly evenly from five to twelve – and the average was seven.

In binoculars, the Pleiades are a stunning sight, while a telescope reveals over two hundred individual stars in the cluster. The brightest stars are blue-white giants, while the fainter stars extend to the cooler colours of the main sequence. On a very dark night, a good telescope will show faint fuzz around some of the stars. Long-exposure photographs reveal streaks of nebulosity that permeate the entire cluster and are brightest where they lie near to the illuminating stars. The streaks show the influence of interstellar magnetic fields that have lined up the gas and dust.

The nebulosity in the Pleiades was for many decades a puzzle. It consists mainly of dust lit up by the stars, and looks totally unlike the glowing gas that survives around clusters of stars that have just been born. The Pleiades are so old, in fact, that their natal gas should have dissipated long ago. The answer became clear in the 1980s, when the infrared satellite IRAS discovered that the Pleiades lie at one end of a cavity within a warm cloud of dust. The young stars are heating the adjacent 'wall' of the cavity. At the same time, the British astrophotographer David Malin produced a superb true-colour photograph of the Pleiades that shows a faint red band of glowing hydrogen adjoining the cluster, and coinciding with the warmest region of dust. The conclusion seems clear: the star cluster has been enveloped by a dark cloud moving from

west to east across the sky, and is lighting up the dust there. This small dust cloud lies at the fringes of the nearest dense interstellar cloud to the Sun, the Taurus Dark Cloud. You can just make out the silhouette of the Taurus Dark Cloud on a very clear night, as a large dark bay (about 10 degrees across) that intrudes into the Milky Way between the Hyades and the Pleiades.

The stars of the Pleiades are not the youngest in our local region. Two striking young clusters – about half the age of the Pleiades – lie in southern hemisphere constellations that once made up the great ship Argo. They are at the same distance as the Pleiades, but in the glowing band of the Milky Way they do not draw attention to themselves in the same way. IC 2391 surrounds the star omicron Velorum, and is a splendid sight in binoculars. IC 2602 is even more spectacular. Known as the 'Southern Pleiades', this cluster is 'bright and splashy even when viewed with the naked eye', in the words of leading Australian observer Gregg Thompson. Its brightest member is theta Carinae, and the cluster appears near the great Carina Nebula in the sky, though it lies at only a fraction of the distance.

With these two young clusters, we have come to the first stars in this chapter that we can describe as genuine children of the Local Arm, rather than elderly visitors just passing through. These juvenile stars did not exist before their raw material entered the region now occupied by the Local Arm, some 30 million years ago. A compression wave was then passing through, and piled up the interstellar matter in this region. The bona fide members of the Local Arm consist of stars up to this age, regions where stars are being born now and dark clouds of gas and dust that have yet to spawn stars.

We can therefore use these young objects to trace out the Local Arm within our neighbourhood. Especially useful are the most massive of the young stars – the O and B spectral types – because they form exceptionally brilliant beacons.

When we plot the location of these young stars, we find

FACING PAGE The IRAS satellite revealed clearly for the first time the extent of the dust cloud that surrounds the Pleiades (the warmest parts are the brightest in this image). However, the Pleiades are too old for this dust to be leftover 'natal' material. Instead, it appears that the dust cloud has drifted past the star cluster, which has cleared out a cavity – the dark region stretching to the left of the Pleiades in this view.

IC 2602, the star cluster nicknamed the 'Southern Pleiades', is less than half the age of its northern counterpart, and its stars were actually born on our doorstep. Young stars like these can be used as excellent tracers of our local spiral arm. Although the cluster is not visually as stunning as the Pleiades, because we see it against the Milky Way in Carina, it contains a number of jewel-bright blue stars.

that they do not lie in the plane of the Milky Way. Instead, they form a flattened system that is 2000 light years across and tipped at an angle of 18 degrees to the Galaxy's plane. This distribution was first investigated by Benjamin Gould, and is known as Gould's Belt (see Box 1). The outer edge of Gould's Belt is marked by the young stars and nebulae of Orion: here the Belt is well below the plane of the Galaxy. Gould's Belt rises appreciably *above* the galactic plane in the opposite direction in our skies, in the region of Ophiuchus, Scorpius and Centaurus.

Gould's Belt is not symmetrical, however. In Scorpius, it is only 150 light years above the plane of the Galaxy, while in the direction of Orion it dips at least 500 light years below the plane. Some astronomers have compared Gould's Belt to a flipper, dangling down from the flat disc of the Milky Way – though it is so small in comparison to the entire extent of our Galaxy that it is more like a fish's scale than a serious appendage.

Because astronomers believe that we live in a typical corner of the Galaxy, structures like Gould's Belt must be common

---

BOX 1. **Discovery of Gould's Belt**

After Galileo discovered that the Milky Way was made of innumerable distant stars, it still struck no-one as odd that some regions of bright stars – in particular, Orion, Centaurus and the head of Scorpius – stood clear of the band of the Milky Way. Even if we live in a flattened star-system, after all, the nearer stars should be spread more widely over the sky than the more distant stars.

The first person to see a global pattern in the brightest stars was John Herschel, who followed up his father's pioneering studies of the northern skies by probing the southern heavens from the Cape of Good Hope. During his stay in South Africa in the 1830s, Herschel noted a 'zone of large stars which is marked out by the brilliant constellation of Orion, the bright stars of Canis Major and almost all the more conspicuous stars of Argo, the Cross, the Centaur, Lupus and Scorpio . . . whose inclination to the galactic circle is about 20°, and whose appearance would lead us to suspect that our nearest neighbours in the sidereal system (if really such) form part of a subordinate sheet or stratum, deviating to that extent from the general mass which seen projected on the heavens forms the Milky Way'.

About the same time, Wilhelm Struve at the Pulkovo Observatory near Leningrad also suggested that our stellar system may consist of two planes of stars, tilted at 10 degrees. The shape of the Galaxy and the Sun's position were unknown at the time, and the discovery of spiral shapes in 'nebulae' led some astronomers to speculate that these two bands of stars were different spiral arms in our Galaxy, with the Sun near the centre of the system.

Without knowing of this previous work – which was far from the mainstream of astronomical research – the American astronomer Benjamin Gould rediscovered the arrangement of bright stars in the 1870s. Like John Herschel, Gould had moved from the northern hemisphere to southern climes, in order to extend the work of mapping the sky. He was director of the Cordoba Observatory in Argentina, and produced the great Cordoba Durchmusterung, a catalogue of 613 953 southern stars that complemented the northern Bonner Durchmusterung of 324 198 stars.

Describing his view of the Milky Way to a north American audience, Gould declared 'in the elevated position and clear atmosphere of Cordoba, this nebulous circle is seen with a vividness far surpassing that to which we are accustomed here . . . And few celestial phenomena are more palpable there than the existence of a stream or belt of bright stars, including Canopus, Sirius and Aldebaran, together with the most brilliant ones in Carina, Puppis, Columba, Canis Major, Orion, &c, and skirting the Milky Way on its preceding side. When the opposite half of the Galaxy came into view, it was almost equally manifest that the same is true there also, the bright stars likewise fringing it on the preceding side, and forming a stream which . . . comprises the constellation Lupus and a great part of Scorpio, and extends onward through Ophiuchus toward Lyra. Thus a great circle or zone of bright stars seems to gird the sky'.

Gould investigated this arrangement of bright stars statistically, and concluded that there was a flattened system of 500 stars that was inclined at 17 degrees to the plane of the Milky Way. He concluded that it was small 'galaxy' superimposed on the much larger Milky Way system. Later astronomers have acknowledged his study by calling the zone of bright stars 'Gould's Belt'.

The first advance on Gould's measurements came when astronomers began classifying the stars by their spectra. Around the turn of the century several astronomers – including Edward Pickering and Harlow Shapley – independently realized that many of the hottest and bluest stars (spectral type B) were members of Gould's Belt. This indicated that the belt forms a rather young system of stars. In the 1950s, Adriaan Blaauw showed that the belt also contains a high proportion of the young star clusters and OB associations near the Sun.

Meanwhile, astronomers cataloguing bright and dark nebulae found that many of these, too were strung along Gould's Belt. More recently, radio surveys have shown that the local interstellar gas – both diffuse hydrogen clouds and small dense clouds containing carbon monoxide – also forms a tilted system that coincides with the plane defined by the Gould's Belt stars.

The American astronomers Richard Stothers and Jay Frogel undertook an exhaustive update of Gould's analysis in the 1970s. They started with a catalogue of 1265 very hot stars (spectral types O to B5), the most luminous supergiant stars and young star clusters, all with accurately known distances. Stothers and Frogel found that these objects clearly form two distinct planes, which intersect near the Sun. About half the objects lie roughly in the plane of the Milky Way, and half in Gould's Belt – which Stothers and Frogel find to be tilted at 18 degrees to the plane of the Galaxy.

When their distribution in space is plotted out, the stars in the Gould Belt system form a dragonfly shape. The two small 'wings' are the Scorpius–Centaurus Association and the stars in Puppis and Vela, while the two longer and more diffuse 'wings' reach from the Sun towards Orion and towards Hercules and Lyra.

Just as astronomers can determine the age of a star-cluster by comparing the spectral types of the stars within it, Stothers and Frogel could work out the age of Gould's Belt by studying the spectra of the stars that make up the distinctive tilted system. The situation is slightly more complicated, because stars are still forming in Gould's Belt, but the method shows that star-formation began in the belt about 30 million years ago.

---

throughout our star system. Gould's Belt appears to be unique just because it is difficult to pick out such a structure unless you are living right in it. But we do see some nebulae and young clusters that lie out of the plane of the Galaxy – the Double Cluster in Perseus (Chapter 4) being a case in point – and these, like Orion, may lie at the tip of a flipper.

These flippers of gas may have been pushed out of the Galaxy's plane by the force of past supernovae. Or they may be portions of the galactic disc that are corrugated, in long-lasting waves that raise ripples in this thin sheet of gas. These vast galaxy waves would travel so slowly that in our short lifetimes they appear to be frozen in place.

The American astronomers Jay Frogel and Richard Stothers suggest that the material in this ripple may oscillate up and down, with a period of 80 million years. In their view, the compression that triggered star-formation 30 million years ago occurred when the material in the flipper passed through the plane of the Galaxy. The previous occasion when the flipper flopped through the Galaxy's plane could have triggered the birth of a previous generation, including the Pleiades – which lies in Gould's Belt even though it is considerably older than the rest of the Belt's stars.

The gas, dust and young stars within Gould's Belt are split up into several separate regions of star-formation. These lie in an oval-shaped ring towards the edge of the belt. According to the Dutch astronomer Adriaan Blaauw, these star nurseries are all related, as daughters of a single burst of star-formation that began 30 million years ago, when the virgin clouds of gas and dust were first compressed to form a spiral arm.

Blaauw has calculated that this compression squeezed a particularly large molecular cloud at a point about 600 light years from the Sun's present position, prompting it to turn into a large cluster of brilliant and powerful stars. This first generation of stars in Gould's Belt would have put the present Orion 'star-factory' literally in the shade. The intense ultraviolet radiation from these first stars, and the powerful blast from their death as supernovae, spread shock waves through the surrounding gas. The force of the shock pushed the gas outwards, to trigger star-birth in a large ring around the edge of Gould's Belt. The jewel in this ring of star-birth is the Orion region; further around are the associations of young stars called Perseus OB2 and Lacerta OB1 (see map in Chapter 5) and the Scorpius–Centaurus Association.

We can see – even with the naked eye – the faded glory of the great starburst that triggered all this activity around us. Look up at the constellation Perseus on a really dark night, and you will notice that the centre of the constellation comprises a spangling of faint stars surrounding the constellation's brightest star, Mirfak. As American amateur astronomer David Eicher enthuses: 'if you scan this area with a pair of binoculars, you'll be delighted at the rich groupings of bright blue-white stars'.

These stars of the Perseus OB3 Association mark the core of the region where Blaauw's starburst occurred. The stars visible in binoculars are main-sequence stars, still burning hydrogen to helium, while Mirfak has gone one stage further. Having consumed all its central hydrogen, it has now expanded to become an orange supergiant. Fainter stars from the great starburst extend into the neighbouring constellations, forming the Cassiopeia–Taurus Association.

Blaauw's idea has been backed up by the Argentine radio astronomer Carlos Olano, who has measured the speed of the hydrogen gas in Gould's Belt by observing its radiation at 21-centimetres wavelength. Olano finds that the gas forms an expanding oval ring 1200 light years across, around the edge of Gould's Belt. Its speed shows that the gas has been swept out from a point practically coincident with the Perseus OB3 Association – and has been moving outwards for about 30 million years.

This ring of expanding gas did not clear all the interstellar matter from the region of Gould's Belt. Astronomers studying radio, optical and infrared wavelengths have in recent years been able to build up a map of how the remaining interstellar matter is arranged near the Sun. It is far from being uniformly spread. The most important discovery is that the interstellar medium is like a Swiss cheese: much of its volume is occupied by 'holes' – or 'bubbles' – typically 500 light years across.

The bubbles are not entirely empty. They simply contain gas that is much more rarefied than average, and is also much

Benjamin Gould (1824–96) realized that the brightest stars in the sky form a 'belt' that is inclined to plane of the Milky Way. As well as the discovery of Gould's Belt, he is remembered for founding the *Astronomical Journal* and for compiling a great catalogue of southern stars while he was director of the Cordoba Observatory in Argentina.

hotter. Astronomers agree that active stars in a young cluster must have blown these bubbles of hot gas in the generally cooler and denser interstellar medium, although they argue about the details. One or more massive young stars in the cluster may 'go supernova' and blast out the cavity. Or the bubble may have been blown by the combined stellar winds of many young stars.

Our local neighbourhood contains at least four bubbles. The Sun, in fact, lies almost exactly in the centre of the Local Bubble – ironically, the most difficult to spot just because it lies all around us. The other three were first found by radio astronomers. The expanding gases have swept up the Galaxy's magnetic fields, so that the outside of the bubble is marked by a shell of more powerful magnetism. Electrons trapped in this field generate synchrotron radiation at radio wavelengths. When we map the radio sky, the nearby interstellar bubbles appear as giant loops of emission stretching up and down from the band of the Milky Way itself.

Even the earliest radio maps of the Galaxy showed part of the most prominent loop, as a long band of emission – the 'North Polar Spur' – stretching from the Milky Way into the northern sky. Later observations extended the spur into a complete ring of emission, which covers almost one-quarter of the sky. If Loop I were visible to the naked eye, it would be an awesome sight, a giant glowing bubble stretching from the south pole of the sky northwards as far as Leo.

Although we cannot see Loop I, the stars responsible for blowing this huge bubble are familiar to every amateur astronomer. They are the stars that make up the body of the scorpion, Scorpius, and most of the stars of Centaurus and Crux, the Southern Cross. The professional researcher knows them more drily as the members of the Scorpius–Centaurus Association.

These stars form an exception to the general rule that the stars in a constellation are unrelated and lie at different distances from the Sun. Most of the stars in Scorpius – including its brightest member, Antares – are around 400 light years away. They are the brightest members of one subgroup of the Scorpius–Centaurus Association, known as Upper-Scorpius.

The stars in the western part of Centaurus, along with almost all the bright stars in the neighbouring constellation of Lupus, lie in Upper-Centaurus–Lupus, the oldest region of the whole association. The third part of the association, Lower-Centaurus–Crux, boasts three of the bright stars in

the Southern Cross, and one of the two 'pointers' – beta Centauri, or Hadar.

The brilliant young stars in the Scorpius–Centaurus Association – the oldest a mere 13 million years – form one of the most prominent parts of Gould's Belt. They lie at the edge of the belt that is nearest to the Sun, and mark the region where it tilts up above the plane of the Galaxy.

Blaauw and his colleagues have traced the history of star-formation in this region. The shock from the original giant starburst in our Local Arm reached this region about 15 million years ago. It compressed the gas on the boundaries of Centaurus and Lupus, to begin the birth of the stars we now see as Upper-Centaurus–Lupus. The massive stars here blew a bubble that squeezed the gas clouds to either side. The effect was felt first to the west, where the stars of Lower-Centaurus–Crux began to form some 12 million years ago. About 5 million years ago, the rash of star-formation spread the other way, to ignite the stars of Upper-Scorpius.

The process is continuing even now. The combined power of all these young stars is blowing up the giant bubble Loop I, and it is squeezing the surrounding clouds of gas and dust to generate more stars. We can see this most clearly in Scorpius. Near to Antares, optical photographs show an elongated dark cloud that is named after rho Ophiuchi, a star just over the border into the constellation Ophiuchus. The expansion of Loop I has compressed this cloud to the point where stars are beginning to condense inside it. The dark dust prevents us from seeing these young stars with optical telescopes, but infrared and radio telescopes have revealed the embryonic stars within.

In the heirarchy of dark molecular clouds, the rho Ophiuchi cloud is fairly insignificant. It contains only one-fifth as much matter as the dark cloud behind the Orion Nebula. Just because it is different from the Orion region, however, the rho Ophiuchi cloud has turned out be highly important for astronomers studying the birth of stars. Whereas the massive and dense Orion cloud is spawning brilliant heavyweight stars, astronomers have found that rho Ophiuchi specializes in making stars that are much lower in mass, more like the Sun and the majority of stars in the Galaxy.

The rho Ophiuchi cloud is not the only dark cloud in the region of Loop I. It is, however, fairly easy to see, because it is silhouetted against quite a bright region of the Milky Way:

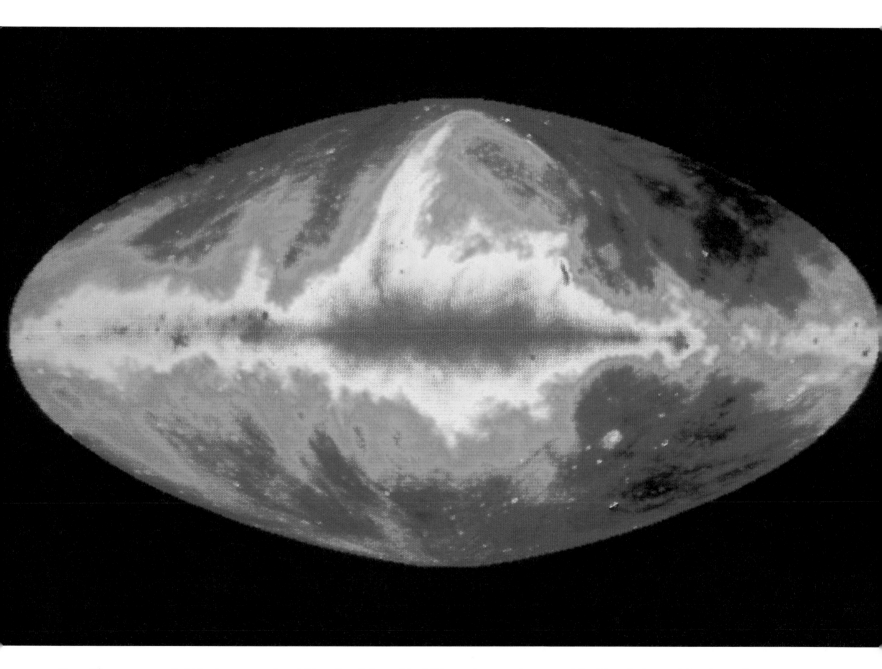

One of four enormous bubbles in our cosmic neighbourhood, Loop I is visible as a huge arc (top centre) in this radio image. The picture shown here covers the whole sky, and was compiled over a period of 15 years using radio telescopes from all around the world. If Loop I emitted light, we would see it spanning the sky all the way from the south pole up to Leo. The stars responsible for blowing this bubble are the brilliant young stars in Scorpius, Centaurus and Crux – the Scorpius–Centaurus Association. Parts of other bubbles are also visible here.

One of the most colourful regions in the sky, the rho Ophiuchi complex is an area of dense dust and glowing gas. The red giant star Antares (lower left) is surrounded by a yellow nebula, while sigma Scorpii (lower right) is cocooned in pink gas. At the top of the image, the star rho Ophiuchi itself is surrounded by dust grains, creating an electric blue glow. At the bottom of the image is the distant globular cluster M4. However, the real site of the action, where star-birth is vigorously underway, is the dark cloud just to the left of centre – where nothing optically can be seen.

FACING PAGE Most of the nearby blue stars in this view towards the centre of the Galaxy belong to the Scorpius–Centaurus Association – a vast group of young stars whose oldest members are a mere 13 million years old. Violent winds from these hot stars are blowing the bubble that radio telescopes reveal as Loop I. In turn, Loop I is compressing the rho Ophiuchi region, and triggering star-birth there.

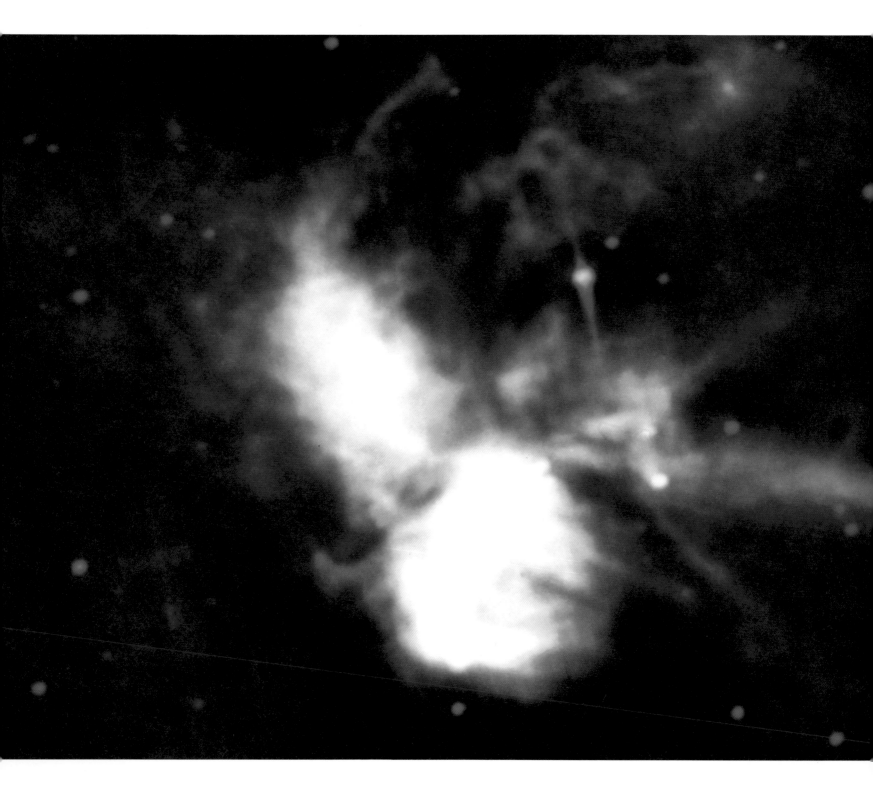

when William Herschel scanned this region of the sky in the 1780s, he was so struck by the lack of stars that he commented 'here is a hole in the heavens'.

Towards the other edge of Loop I is a dark cloud that is even more prominent, because it lies against the main band of the Milky Way. The Coal Sack is unmistakeable, even at the most casual glance – a dark hole in the middle of the Milky Way, right next to the Southern Cross. The Aborigines of Australia – accustomed to skies so black that they made constellations from dark patches rather than stars – saw the Coal Sack as an emu lying in wait for a possum perched in a tree represented by Crux. Early European navigators saw it as the antithesis of the Magellanic Clouds, and called it Magellan's Spot.

In fact, the Coal Sack is fairly tenuous as dark clouds go. Containing enough material to make no more than 40 000 suns, it is only one-tenth as massive as the rho Ophiuchi cloud – and this gas and dust is spread over 60 light years. As a result, powerful telescopes can in fact see right through parts of the Coal Sack.

Photographs show several other dark clouds in this region. Where clouds are not silhouetted against a starry background, we can track them down by their emission at radio wavelengths. The group from Columbia University who pioneered the mapping of molecules in the Milky Way (Chapter 3) have found the shapes and distances of clouds in this region that lie out of the plane of the Milky Way – possibly pushed there by the expansion of Loop I.

On the map, many of these clouds seem to lie inside Loop I. In fact, there are sufficient uncertainties both in their distances and in the shape of Loop I where it intersects the Galaxy's plane that all these dark clouds could well lie on the edge of Loop I, squeezed by its expansion. The dark clouds form part of Olano's expanding hydrogen ring at the edge of Gould's Belt. They are also part of the lane of dust that lies along the inner edge of our Local Arm, stretching from the dust clouds of Vela to the great dark rift in Cygnus (see previous chapter). Viewed from outside, we would see that the burst of star-formation in Scorpius–Centaurus has ripped this portion of the dust lane to shreds.

Rho Ophiuchi marks one end of the intact dust lane, heading through the Aquila Rift and out of the local neighbourhood towards Cygnus. Soon, however, the Aquila Rift may find itself under pressure from another side, from Loop III. This bubble shows up quite clearly on radio maps, projecting from the northern side of the Milky Way like a smaller and fainter version of Loop I. Its existence is something of a mystery, however, as there is no association of powerful stars to drive its expansion. Possibly it is the expanding shell from one single supernova that exploded around a million years ago.

Within Loop III are some stars that have become prototypes for various classes of astronomical objects, just because they are the nearest to the Sun. RR Lyrae is an elderly star that varies regularly in size – and brightness – with a period of 14 hours. Stars of this type are important in measuring distances, because all RR Lyrae stars have the same intrinsic brightness: averaged over a cycle, an RR Lyrae star is about 50 times more brilliant than the Sun.

SS Cygni is also a variable star, but of a very different type. This star brightens suddenly every couple of months for a short while, rather like a nova explosion but on a smaller scale and much more frequently. From studies of SS Cygni and other similar 'dwarf novae', astronomers have deduced that it consists of a white dwarf drawing material from a companion star. The infalling matter forms an accretion disc around the white dwarf. When the disc accumulates enough gas, it becomes unstable and flares up to a hundred times its previous brightness. In its diminutive outburst, the dwarf nova does not eject gas into space.

Another double star, in the same region, makes quite a show about ejecting gas. R Aquarii puzzled astronomers for a long time because its spectrum showed that it is simultaneously a hot star and a cool star. This 'symbiotic star' is in fact a cool red giant star and a small hot companion in close orbit. Gas falls from the large star towards the companion. Every 40 years or so, however, R Aquarii spews some of this matter out into space. The 'bullets' of gas form a jet that currently stretches for 20 times the width of the Solar System. R Aquarii forms the nearest example of an astrophysical jet, and astronomers

FACING PAGE In the infrared, the rho Ophiuchi cloud glows brilliantly – quite the opposite of its optical appearance, described by William Herschel as 'a hole in the heavens'. Unlike larger complexes such as the Orion Nebula, the rho Ophiuchi region specializes in making average stars like the Sun instead of giant stars.

Silhouetted against the stars of the Milky Way, the Coal Sack is a dark cloud of dust and gas that is easily visible to the unaided eye next to the Southern Cross. Although dramatic in appearance – because of its contrast against the Galaxy's star clouds – the Coal Sack is only one-tenth as massive as the rho Ophiuchi cloud. Nevertheless, it contains enough material to make 40 000 stars.

hope it will provide clues to the nature of the much more powerful jets in regions of star-birth and in quasars.

Adjoining Loop III is the slightly more prominent Loop II, an interstellar bubble projecting to the south of the Milky Way. Within Loop II lies Mirfak and the Perseus OB3 Association, which was probably responsible for the huge ring of expanding gas that has triggered star-formation throughout Gould's Belt. More recent activity here – possibly just a single powerful supernova – has now blown the smaller Loop II bubble.

Again, we find some prototype stars within the loop. Towards the far edge is gamma Cassiopeiae, a bright blue star ten times the size of the Sun that is spinning 200 times faster than our star. Its rotation is throwing shells of gas from around its equator, causing the star to fluctuate irregularly in brightness. Long-exposure photographs show odd wisps of nebulosity around the star.

At the northern edge of Loop II is Polaris. As well as guiding navigators, the Pole Star is one of the nearest examples of a Cepheid variable (the prototype, delta Cephei, lies just off this map) – or rather, it *was*. The star has always had relatively small changes in brightness, and in the latter half of the twentieth century its pulsations have been getting weaker. Theory suggests this should happen as a star grows older, and the end of Polaris's phase as a Cepheid variable is a rare example of stellar evolution occurring within a human lifetime.

Zeta Aurigae, at the eastern edge of Loop II, is one of the most unusual eclipsing binary stars that we know. It consists of a cool supergiant star 150 times wider than the Sun and a small hot main-sequence star. This odd couple orbit one another in a period of 2.7 years. For over a month in each period, the supergiant hides the smaller star from us. This star is the prototype of half-a-dozen known zeta Aurigae stars, one of which – by pure concidence – is epsilon Aurigae, its next-door neighbour in our skies but in fact a totally unrelated star several times further away (see Box 3 in Chapter 5).

This edge of Loop II is currently squeezing a patch of interstellar material, and provoking an outbreak of star-birth. The compressed material appears in our skies as the streaks of dark matter that comprise the Taurus Dark Cloud – the cloud that lies near to the Pleiades. Much more important to astronomers, the Taurus Dark Cloud is the nearest region of star formation to the Sun, only 450 light years away.

The Taurus Cloud is relatively flimsy as compared with its neighbour in the sky, Orion. So – like the rho Ophiuchi cloud – it is forming relatively lightweight stars. It is the ideal

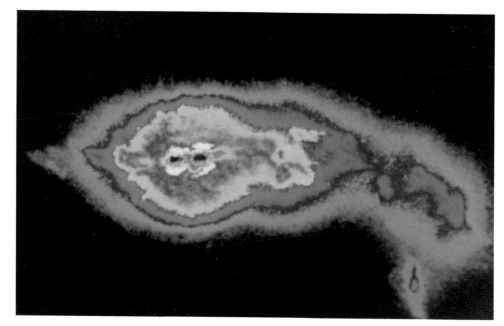

Hubble Space Telescope image of the peculiar star R Aquarii, which is probably a cool red giant star circled closely by a white dwarf. The two dots at the centre of the image are the star-system itself, and the surrounding structure is gas that has been erupted out of the system. The interaction between the two stars has ejected gas streamers out to a distance of 20 times the diameter of the Solar System.

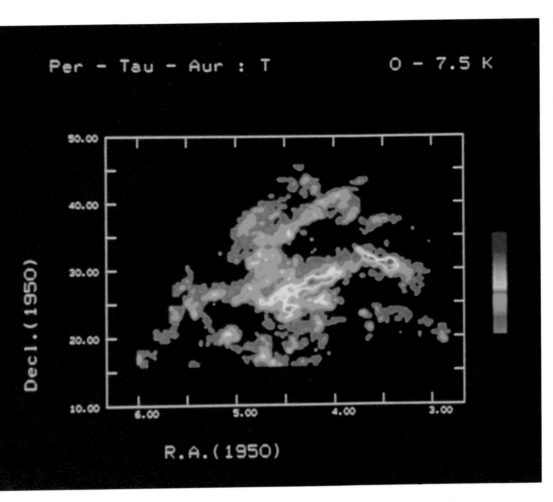

Per - Tau - Aur : T   O - 7.5 K

The Taurus Dark Cloud, 450 light years away, is the nearest molecular complex to the Solar System. This false-colour map shows the intensity of carbon monoxide in the cloud, which is a good guide to its density. Unlike giant molecular clouds such as Orion, the Taurus Dark Cloud is collapsing to form small stars like the Sun. The dense knots in the cloud (coloured yellow and red) show where young stars are starting to form (except for the very bright region at right, which is actually the distant Perseus OB2 Association, lying in the line-of-sight).

place for astronomers to study the birth of stars like the Sun.

Radio astronomy provides the tool for investigating the first stages in this process. The emission from carbon monoxide shows how the cloud has fragmented into individual sites for star-birth – rather different from the monolithic cloud in Orion where massive stars are made. To pin down what happens when the gas inside a clump becomes very dense, we must switch to another molecule. By studying emission from ammonia molecules, the American astronomer Philip Myers has discovered many 'dense cores' in the Taurus clouds. These are small regions where the gas is so densely concentrated that its own gravitational influence must lead to an inexorable collapse to become a star. This is the stage immediately before star-birth.

Some of the dense cores contain a source of infrared radiation, discovered in the pioneeering survey by the infrared satellite IRAS. Here, we are seeing a star embryo embedded in the still-collapsing core. The dust in the surrounding dense core dims the optical light by 30 magnitudes – amounting in practice to invisibility. Their infrared radiation, however, shows that these young stars have about the same mass as the Sun, but are 10 to 20 times brighter than our star. In these youngsters, the energy from nuclear reactions at the centre is being augmented by heat released by gas and dust still falling onto the star, and by the shrinkage of the star itself as it settles down to its final size.

In front of the Taurus Dark Cloud, ordinary telescopes can

BOX 2. A smoky chimney

Astronomers have long known that interstellar space is polluted. It's full of small dust grains that are mainly tiny blobs of rock and 'soot' condensed from the outer layers of old red giants as they expand to become planetary nebulae. In the process, atoms of carbon go through some complex chemical reactions. This phase of a star's life is relatively short, so it is difficult to catch a star in the act of polluting its neighbourhood; but we are lucky because one such star lies quite near to the Sun.

CW Leonis was first catalogued as a rather dim variable star – a pulsating red giant star that was shrouded in dense dust. When American astronomers Gerry Neugebauer and Bob Leighton scanned the sky for the first time at infrared wavelengths, in the 1960s, this star stood out much more prominently by its emission at a wavelength of 2.2 micrometres. This pioneering Infrared Catalogue gave the star its often-used – but cumbersome – designation IRC + 10°216. At a wavelength of 5 micrometres, CW Leonis is the brightest object in the sky – the 'Sirius' of infrared astronomy.

In fact, CW Leonis is an intrinsically very luminous red giant, and without its cloak of dust it would be one of the brightest stars that we would see in the sky. As it is, the surrounding dust absorbs most of the star's light; as a result, the dust is heated to a temperature of 400 degrees Celsius and then radiates the energy away again as infrared.

CW Leonis is also a treasure chest for astronomers seeking out molecules: it contains at least 40 different chemical compounds, including the largest molecule ever detected in space – a string of 13 atoms with the formula $HC_{11}N$. These molecules have, in the main, been identified by their emission at specific wavelengths in the radio region of the spectrum.

The existence of this smoky chimney right on our doorstep – about 600 light years away – has given researchers a unique chance to study chemical reactions in a situation very different from a chemical laboratory on the Earth – indeed, many of the molecules in CW Leonis have not yet been synthesized in the laboratory. Among the species found around this star are long chains of carbon atoms linked by a variety of chemical bonds, molecules where three atoms – of carbon or silicon – form a triangle, and even sodium chloride – common salt!

Almost certainly, some of the carbon atoms around CW Leonis have condensed into 'buckyballs': spherical molecules of 60 atoms each, where the bonds form the edges of hexagons and pentagons like the pattern on a football or a geodesic dome. Their official name is buckminsterfullerene, as a tribute to the American architect Buckminster Fuller who designed the first buildings in the form of geodesic domes.

Near to the star, the outflowing carbon-rich gas condenses into molecules and larger dust grains. As they flow out from the star, however, the molecules are subjected to ultraviolet light from other stars, which sets off another chain of chemical reactions that breaks the large molecules apart again. In the end, the old star will become a planetary nebula (see previous chapter), and the central white dwarf will emit intense ultraviolet radiation and a fast wind of hot gases. These disruptive forces will destroy – in only 100 years – all the molecules that have been built up in the 20 000-year-long chemistry experiment that we call CW Leonis.

---

pick out about 100 stars that have recently emerged from these dark cocoons. They lie in loose clusters, each a few light years across and often associated with a few dark cores where members of the cluster are still being born. While the massive stars in Orion have more or less settled down to maturity by the time they emerge, the low-mass Taurus stars are far from stable when we first see them.

The prototype of these young Sun-like stars is T Tauri. This star flickers erratically, and it illuminates a nebula that also changes in brightness – a phenomenon first noted by the Victorian astronomer John Russell Hind, giving rise to the name 'Hind's Variable Nebula'. The light from T Tauri varies partly because it is surrounded by shifting clouds of dust, but also because the star's outer layers are unstable. The star is still shrinking to its final size, giving rise to giant convection currents in its outer layers that constantly change the star's luminosity and tangle up the surface magnetic field to trigger brilliant flares.

The powerful convection currents also energize the corona (the outer atmosphere) of T Tauri. A powerful wind of particles blows outwards from the corona of T Tauri, stripping away any gas and dust that has not either fallen onto the star or condensed into planets. Astronomers believe that such a 'T Tauri wind' was responsible for clearing our Solar System of stray gas and dust soon after the birth of the Sun and planets.

While T Tauri provides a glimpse of the Sun as it was about a million years after its birth, one of its neighbours – HL Tauri – shows how both the Sun and the Solar System would have looked at only one-tenth this age. Ten per cent of the T Tauri stars – including the prototype – are surrounded by dust, and in the case of HL Tauri, this dust is clearly marshalled into a disc.

According to British astronomer Martin Cohen, the disc surrounding HL Tauri is about three times the size of the Solar System. The inner part contains dust that is composed of rocky grains, while the outer part – beyond the radius of Saturn's orbit – contains icy particles as well. This is just the structure (albeit on a slightly larger scale) that Solar System theorists need in order to create rocky planets near the Sun and gaseous worlds further out. With new telescopes coming on stream all the time, astronomers are confident that the dark clouds of Taurus will eventually reveal to us in detail just how our own planetary system was born.

The last of the four bubbles in our local neighbourhood is the smallest, and we would perhaps know nothing about it if the Sun did not happen to lie right in the middle of it. In itself, the Sun's location does not help, because we suffer the cosmic analogy of not being able to see the wood for the trees: the Local Bubble lies all around us, so its gas appears in every direction that we look. But the extent and content of the Local Bubble has been established by several lines of investigation, in particular by astronomers who have checked the spectra of hundreds of nearby stars to look for narrow absorption lines caused by interstellar gas between the star and us.

Donald Cox and Ronald Reynolds, of the University of Wisconsin, have collated these measurements to build up a consistent picture of the Local Bubble. They find that it is about 300 light years across in the plane of the Milky Way. The bubble is far from spherical, however: it stretches further out of the plane where the density of the surrounding gas is lower, and so has an overall shape rather like a barrel.

The walls of the Local Bubble are marked by regions of denser gas. In the directions of Aquila and Cassiopeia–Perseus, they abut directly onto relatively high-density regions beyond, while the walls of the Local Bubble probably merge with those of Loops I and III in the directions of Scorpius and Cygnus – like adjacent cells of a honeycomb. In the direction of Puppis and Canis Major, however, the wall is relatively thin and there is little gas or dust beyond. By coincidence, the outer arms of the Galaxy are also fairly tenuous in this direction, so this 'Puppis Window' in the band of the Milky Way gives astronomers an unusually clear view of distant stars and of the Universe beyond.

The interior of the Local Bubble is remarkable empty of matter: the density of gas here is only one-twentieth the average for the Galaxy. Despite its low density, astronomers have some direct measurements of its properties. The thin gas is very hot, with a temperature of around a million degrees, and it emits low-energy X-rays copiously. This gives rise to a background of 'soft' X-rays over the entire sky.

Astronomers are still arguing about the origin of the Local Bubble. It may be a still-expanding supernova remnant, blown up by a star that exploded only 50 to 100 light years away from the Sun, about 100 000 years ago. Such an explosion could explain the presence of the short-lived isotope aluminium-26 in the nearby interstellar gas, revealed by its characteristic emission of gamma rays. The collapsed core of this star may be a strange gamma-ray-emitting pulsar called Geminga (Box 3). But the chances are very slim that the Sun would happen to lie inside such a young remnant. The alternative is that the Local Bubble is a 'fossilized remnant' – a cavity blown by a supernova over a million years ago. The bubble reached its final size soon after the explosion, but it has preserved its size and shape because the pressure of the interior hot gas has kept it inflated ever since.

Within the Local Bubble, we find several small clouds of slightly denser gas. Astronomers have mapped these clouds by investigating in even more detail the light from stars in our vicinity, looking both for dark spectral lines etched by the gas in space and for polarization of the starlight that is caused by interstellar dust grains lined up in the Galaxy's magnetic field. These measurements reveal several elongated clouds lying between the Sun and the wall of denser gas that the Local Bubble shares with the giant Loop I. These clouds are probably outlying streamers of gas from Loop I.

This gamma-ray view of Gemini and Taurus, taken by the EGRET telescope on NASA's Compton Observatory, shows the Crab Nebula (centre), and Geminga (pink blob). Geminga is the second-brightest gamma-ray source in the sky, but its optical counterpart is one of the faintest stars known. It is the closest neutron star to the Sun, lying about 100 light years away. The surface temperature of Geminga is estimated to be 500 000 degrees Celsius.

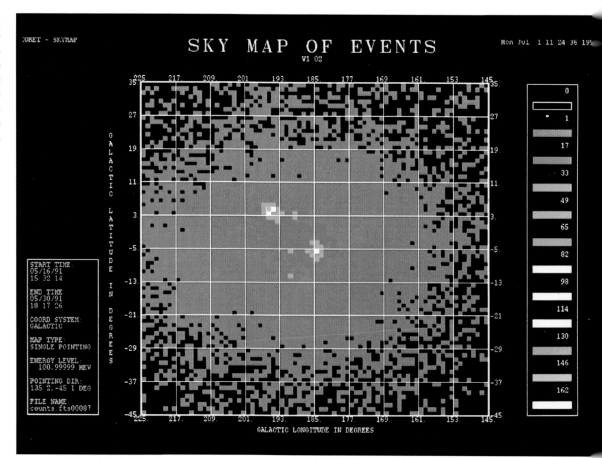

BOX 3. **The gamma-ray source that 'does not exist'**

Geminga is the name of an object that has puzzled astronomers for two decades. The name is an abbreviation of 'Gemini gamma-ray source', but it also – with good reason – means 'does not exist' in Milanese dialect!

In 1972, the American satellite SAS-2 opened up the field of gamma-ray astronomy by making the first survey of the sky at these wavelengths. It found strong emission from the Milky Way in general, and from the two young pulsars in the Crab Nebula and in Vela. It also found gamma rays coming from Gemini.

The European satellite COS-B gave astronomers a vastly improved view of the gamma-ray sky, including a catalogue of 25 individual sources. Using data from this satellite, the Italian astronomer Giovanni Bignami found that the source in Gemini was the second-brightest in the gamma-ray sky – after the Vela Pulsar – and he could pin down its position to within a fraction of a degree. To his surprise, optical photographs showed no unusual object there, and existing catalogues of radio and X-ray sources contained nothing at this position. At this point, Bignami suggested the name 'Geminga' for the gamma-ray source that did not seem to exist at any other wavelength.

But some determined sleuthing eventually unearthed other signs of the elusive gamma-ray source. In 1981, Bignami and his colleagues searched the vicinity of the gamma-ray source with the X-ray telescope on the orbiting Einstein Observatory. This revealed an unusual X-ray source that was almost certainly Geminga – and gave a much more precise position for it.

Armed with this new location, optical astronomers searched once more. Unfortunately, Gemini lies right in the Milky Way, and any one of dozens of faint stars could coincide with the X-ray position, even given its relatively small uncertainty. In 1987, however, American astronomers Jules Halpern and David Tytler found that one of the faintest detectable stars in the region (with a magnitude of 25) was unusually blue, indicating a temperature at least ten times higher than any normal star. Putting the optical and X-ray data together, Halpern and Tytler deduced a temperature of around 500 000 degrees Celsius for Geminga.

The next clue came from Rosat, a German X-ray telescope launched in 1990. The following year, Halpern used Rosat's sensitive instruments to investigate the X-rays from Geminga. He found that the radiation is not constant, but pulses regularly with a period of just under a quarter of a second. This discovery at last linked Geminga with a known class of astronomical object: the pulsars.

A pulsar is a compact neutron star, spinning about once a second. It emits radiation in a beam, rather like a light-house, and as it spins around we pick up a 'pulse' of radiation once per rotation. Most pulsars are best at producing radio waves, but the youngest pulsars also emit light, X-rays and gamma rays. The prototypes of these youngsters are the Crab and Vela pulsars, with ages of 950 and 12 000 years. Geminga, it seemed, must also be a comparatively young pulsar.

With the Rosat discovery, the gamma-ray trail livened up again. Gamma rays from a pulsar carry a lot of energy, but they arrive only infrequently, often missing out thousands of pulses. With such large gaps, it is impossible to work out a period from the gamma rays alone. Armed with the precise period from Rosat, however, Bignami could check whether the individual gamma rays – detected by satellites from COS-B in 1975 to NASA's Compton Gamma Ray Observatory in 1991 – arrived in time with the X-ray pulses. The answer was a resounding yes.

Looking back over these years, however, Bignami found that slight discrepancies began to creep in. The pulsar, he concluded, is gradually slowing down. This was no great surprise, because both the Crab and Vela pulsars are slowing down as they lose rotational energy. Most important, the rate of slowing depends on the pulsar's age. Geminga, Bignami calculates, is the relict of a star that exploded 370 000 years ago.

By comparing Geminga with the Vela pulsar, Bignami has also worked out its distance. At about 100 light years, it turns out to be the closest pulsar to the Earth. Geminga appears in our skies as a surprisingly strong gamma-ray source only because it happens to lie much closer than any other neutron star.

Even so, no-one could claim that all the mysteries of Geminga have been solved. In particular, radio astronomers have still found no trace of it. A neutron star with the power to generate high-energy gamma rays should also speed up electrons to high velocities, and these electrons would broadcast radio waves by the synchrotron process.

Geminga is almost certainly the prototype of many objects in the Galaxy. In fact, many of the weaker gamma ray sources found by COS-B and the Compton Gamma Ray Observatory also lack obvious counterparts at other wavelengths. They are probably more distant versions of Geminga – the source that now definitely 'does exist'.

The Sun, it turns out, lies within a distinct small cloud, the Local Fluff. We are near one end of this elongated cloud, which stretches some 25 light years in the direction of Cygnus, and is 5 to 10 light years wide. The density of the gas and dust in the Local Fluff is about 10 times higher than the tenuous surrounding gas in the Local Bubble: but, even so, the Local Fluff does not contain enough matter to make up a star like the Sun.

Because the Sun is embedded in the Local Fluff, its gas and dust should permeate the Solar System, and we should – in principle – be able to sample this interstellar material directly. But the Sun itself causes a major problem. The fast-flowing solar wind, streaming out from the Sun's surface, blows its own tiny bubble in the interstellar medium. This heliosphere keeps out ionized gases from beyond. Some neutral atoms of hydrogen and helium do creep in, however. Astronomers have investigated these atoms by the sunlight that they reflect – mainly at ultraviolet wavelengths – using rockets, Earth-satellites and interplanetary spacecraft. They find that this 'very local interstellar medium' is at a temperature of 10 000 degrees Celsius and has a density of about 0.1 atoms per cubic centimetre – similar to the values deduced for the Local Fluff, and about the average for the gas in the Galaxy's disc as a whole.

The gas is blowing through the Solar System, in the form of the 'local interstellar wind', at a speed of 25 kilometres per second (about twice the speed of a rocket escaping from the Earth's gravity) from the direction of Ophiuchus. Part of this measured velocity is caused by the fact that the Sun and the Solar System are themselves moving (see Box 4 in Chapter 3). We need to subtract the Sun's velocity to find the true speed and direction of the local interstellar wind, just as a sailor judging the wind must allow for the ship's own motion. According to Priscilla Frisch and Donald York, at the University of Chicago, the local interstellar wind is actually blowing at a speed of 20 kilometres per second from the general direction of alpha and beta Centauri. They conclude 'this flow appears to be driven by the relic superbubble Loop I'.

The bearing of the local interstellar wind shows that the elongated Local Fluff is moving sideways-on. It is almost certainly one of the first wisps of gas propelled towards us by the expanding Loop I, the vanguard of its advance as it begins to crush our smaller bubble. From the dimensions of the Local Fluff, and its motion, astronomers have calculated that the Sun is just entering this wisp of gas and dust, and will take about 200 000 years to pass through.

At present, we are in the lower-density fringes of the Local Fluff, but its central regions undoubtedly contain some denser cloud cores. Frisch and York warn: 'upon encountering the Solar System, these cloud cores will suppress the solar wind, lead to the accretion of interstellar material onto the solar surface, and perturb the equilibrium of the terrestrial oxidising atmosphere enough to alter the climate on Earth'.

Turning to the stars that lie within the Local Bubble, we find nothing that rates on the galactic scale: the brightest star here, Alkaid, is as luminous as 300 Suns – a glow-worm in comparison to the brilliant beacons that we find, for example, in the Scorpius–Centaurus Association. This is no surprise, really, because the Local Bubble is so small that it is unlikely to contain any of the rather rare highly luminous stars. In the main map in this chapter (Map A, p. 158), we have plotted only stars brighter than 250 Suns, and only four lie within the Local Bubble: Achernar, Alkaid, Elnath and Scheat.

Just to put our own star in perspective, note that the larger-scale map of the Local Bubble itself (Map C, p. 160) shows the 50 or so brightest stars in the region – and even the faintest of these is around a hundred times brighter than the Sun. There are literally thousands of stars within the Local Bubble alone that outshine our star. To illustrate the point, we have marked on Map C an obscure star in Puppis called HD 44594, which is just too faint to be seen by the naked eye. HD 44594 is in fact the twin of the Sun – the nearest match to our star that astronomers have been able to pin down. Just as we do not rate HD 44594 highly, someone in this star's planetary system – if it has one – would regard our Sun as a totally insignificant star, invisible without a telescope, even though it lies only 80 light years away.

On the large-scale Map C, we have also marked the first-magnitude stars that are familiar from our night sky. But such star-sights as Sirius and Vega only appear glorious to us because they lie nearby: on a map that strictly showed only the 50 most luminous stars in the Local Bubble they would not appear at all. So the Local Bubble is really the Galaxy's equivalent of life in an everyday suburb – away from the hot-spots and violence of downtown.

The Local Bubble is, however, a great fishing ground for double stars. Many of the stars in the Galaxy must in reality be double systems, but the constituent stars are so close together that we cannot easily see them as separate objects. By investigating the stars close to us, astronomers can tell how common double stars actually are.

The German astronomer Wilhelm Gliese has made a detailed study of the nearest stars to the Sun. Within 16 light years of the Sun – the region where we probably have a complete inventory, even of the faintest stars – he lists 27 single stars, 14 double stars and five triple systems. This means that if we pick something that looks like 'a star' at random in the sky or on a photograph, there is a 40 per cent chance that it will actually be a double or multiple system.

The nearest star-system to the Sun, alpha Centauri (left) is a triple star. It consists of two components rather like the Sun itself, and a red dwarf, Proxima Centauri, which is the closest-known star to our Solar System. Beta Centauri (centre) – also known as Hadar – is a much more remote blue-white giant star more than 10 000 times brighter than alpha.

FACING PAGE These peculiar 'comet-like' stars were imaged by moving the lens focus about twelve steps over the period of a half-hour exposure. This technique reveals the colours of the stars well: note the whiteness of alpha Centauri (far left), compared to the blueness of beta (right of alpha). The stars in the Southern Cross (right) are also predominently blue, with the glaring exception of red gamma Crucis.

We can also look at these statistics another way. The double and multiple systems within 16 light years actually contain 43 stars (28 plus 15), so outnumbering the single stars. So it's fairer to say that the majority of stars in the Galaxy (about 60 per cent) lie in binary or more complex systems.

When it comes to stars on our doorstep, these double and multiple stars provide a feast for anyone with a small or moderate telescope. A case in point is epsilon Lyrae, which lies near Vega in the sky but is in fact considerably more distant. With very good eyesight, you can just see that this star is double, while a telescope reveals that each of these components consists of two stars, giving epsilon Lyrae the nickname the 'Double-Double'. The American astronomer and historian Richard Allen called it 'by far the finest object of the kind in all the heavens'.

Double stars can also provide a glorious colour contrast: the natural differences in star colours are often accentuated when stars of different hues are seen close together. Not far from epsilon Lyrae in the sky, and lying just outside the Local Bubble, is Albireo: even the smallest telescope shows it to be a bright yellow star with a blue companion. The Victorian astronomer Agnes Clerke described the components as 'golden and azure . . . perhaps the most lovely effect of colour in the heavens' – this was before modern photographic techniques allowed astronomers to bring out vivid colours in nebulae that cannot be seen by eye!

In the same region of sky, but slightly closer, is the triple system Almach – described by Richard Allen as 'orange, emerald, and blue . . . the contrast in their colours is extraordinarily fine'. Triple stars and double-doubles are by no means the end of the story. Castor is one of the most remarkable systems known: an inhabitant of a planet there would be treated to no fewer than six suns! A small telescope shows that the single star we see with the naked eye is in fact double, with a fainter star orbiting them much further out. By splitting the light from these stars with a spectroscope, astronomers have found that each of these three stars is in fact a close double.

The spectroscope is not the only way to resolve a double that cannot be split by a telescope. The two main stars in the Algol system are too close to be resolved with any telescope – yet naked-eye observations revealed that this must be a double star. The system happens to be tilted so that the two stars pass in front of one another as they pursue their orbits.

When the brighter of the two is hidden, the total light from Algol drops to only one-third of its normal brightness. This behaviour was discovered, and interpreted correctly, in the eighteenth century by a remarkable young English astronomer, John Goodricke, who was both deaf and dumb.

The companion to Sirius – the Dog Star – was also 'discovered' before it was ever seen or photographed. In the 1830s, the German astronomer Friedrich Bessel found that Sirius was not moving in a straight line through space, and in 1844 he predicted that Sirius was moving about the centre of gravity of the star itself and an unseen companion. This faint star was first seen nearly two decades later, by the American astronomer Alvan G. Clark who was testing a new and powerful telescope. Bessel also predicted, correctly, that Procyon – coincidentally the brightest star in the constellation of the Little Dog – has a faint companion.

Early in the twentieth century, astronomers found that these companion stars were every bit as hot as ordinary stars: because they are thousands of times fainter, however, they must be much smaller than normal stars. The companions to the two dog stars are only about the size of the planet Earth, and the combination of their colour and diminutive dimensions has led to the name 'white dwarf'. Despite their small sizes, however, white dwarfs are not particularly low in mass. By studying the orbits of stars like Sirius's 'Pup', astronomers have been able to deduce the masses of dozens of white dwarfs, and most weigh in at about two-thirds the Sun's mass. This means that the matter in a white dwarf is compressed to a million times the density of water.

Although this density seemed astonishingly high to astronomers at the turn of the century, physicists found a natural explanation in the 1920s, with the development of the new quantum theory to explain the structure of matter. The high temperatures and pressures in a white dwarf not only break apart atoms, but also pack the components – electrons and atomic nuclei – as close together as is possible, in what physicists call a degenerate state. According to quantum theory, the electrons effectively have a larger size than the nuclei, because they are lower in mass, and it is the pressure of electrons pushed shoulder to shoulder that determines the size of the white dwarf.

Degenerate matter is easily squashed by extra gravitational force, and this means that a more massive white dwarf is smaller than a lightweight white dwarf. In the early 1930s, the Indian astrophysicist Subrahmanyan Chandrasekhar calculated that a white dwarf as heavy as 1.44 solar masses would have zero size – and this maximum mass for white dwarfs is known as the Chandrasekhar limit. Chandrasekhar was awarded the Nobel Prize for this work – over half a century later!

Because white dwarfs are the 'corpses' of the relatively abundant low-mass stars, they must be very common in the Galaxy. There is probably one white dwarf for every ten ordinary stars. But white dwarfs are so faint that we can only study those that lie within a few hundred light years of the Sun. Most of our information has come from white dwarfs that have been picked out because they are companions to normal bright stars – like Sirius and Procyon. White dwarfs are much hotter than most other stars, however, with temperatures up to 100 000 degrees Celsius, and they produce copious amounts of radiation at extreme ultraviolet wavelengths. The pioneering survey of the sky at these wavelengths, by the Rosat orbiting observatory in 1991, picked up hundreds of previously unknown white dwarfs in the Sun's vicinity.

Also faint and difficult to observe are the faintest stars on the main sequence. They are popularly known as 'red dwarfs', because they have low temperatures and are only a fraction of the Sun's size. The name is rather confusing, though. Red dwarfs have no resemblance to white dwarfs: they consist of normal gas (rather than degenerate material) and hydrogen is fusing to helium in their cores. They are, in fact, merely smaller and slower-burning versions of the Sun. The smouldering fires of a red dwarf can last so long that the star typically has a lifetime of 100 000 million years – far longer than the current age of the Universe.

Red dwarfs are by far the commonest kind of star in the Galaxy. When a molecular cloud condenses into stars, it forms comparatively few heavyweight stars, many mediumweights like the Sun and a profusion of the lowest-mass stars. The inventory of stars within 16 light years of the Sun reveals that two-thirds are red dwarfs. It's no coincidence, then, that the nearest star to the Sun is a red dwarf: Proxima Centauri, lying 4.24 light years away. Wilhelm Gliese has calculated that another red dwarf, Gl 710, currently some 50 light years away, is heading so directly at the Sun that in 650 000 years' time it may pass only one and a half light years from the Solar System.

The dimly glowing red dwarfs are the faintest of the main sequence stars: they are typically only one-thousandth as bright as our Sun – although many red dwarfs are hotbeds of magnetic activity, and can erupt in flares that shine hundreds of times more brightly than the star itself. Among the contenders for 'dimmest star' are VB 10, which lies 18 light years away in the direction of Aquila, and RGO 0050-2722 in Sculptor. At a distance of a 'mere' 80 light years, the latter star is invisible to all but the largest telescopes. Both these stars are around a million times fainter than the Sun. If we put VB 10 or RGO 0050-2722 in the centre of our Solar System, they would illuminate our skies no more brightly than the Full Moon.

These two dim red dwarfs are the lowest-mass objects that we can call a star – an object that shines because it contains a nuclear fusion reactor. We have no direct measurement of their masses, but calculations tell us that a ball of gas can only start to fuse hydrogen to helium if its mass is more than one-twelfth the mass of the Sun – equivalent to 80 times the mass of the planet Jupiter.

An object above this limit is a red dwarf; but what do we call something less massive than 80 Jupiters? Until the 1980s, astronomers used the term 'planet', because the only objects we knew in this mass range were the Sun's family of small worlds. In addition, stars are hot because they 'burn' nuclear fuels, and a naive analysis would suggest that anything below this limit would be cold, and emit no light or radiation of its own – again like one of the Sun's planets.

But new theories of planet formation now suggest that the Sun and its planets are not just different in mass: they formed in completely different ways. The Sun collapsed from a large cloud of gas, while the planets built up from small pieces of solid rubble that were a by-product of the birth of the Sun. In regions where stars are forming, we should expect collapsed fragments of gas covering a wide range of masses, even below the limit for hydrogen burning. These objects – with masses between about 10 and 80 Jupiters – are much more akin to stars than they are to planets.

Faced with a new kind of object, astronomers had to dream up a new name: 'brown dwarfs'. On the stellar scale, they are certainly dwarfs, about one-tenth the diameter of the Sun. 'Brown' is a somewhat tongue-in-cheek term: these objects do not emit any noticeable quantity of light, but they are not totally 'black' because they do radiate at infrared wavelengths, from a surface that is at a temperature of about 1500 degrees Celsius. Although a brown dwarf has no long-lasting source of energy, it is born warm. This energy comes partly from the contraction of its gases under the pull of its own gravity, and partly from a short-lived burst of nuclear energy as the tiny amount of deuterium – heavy hydrogen – that it contains fuses into helium.

With such a solid theoretical basis behind them, observational astronomers have begun to look in earnest for the real thing. They have been helped by the development of infrared cameras that can pick out brown dwarfs by their predicted weak output of heat radiation. Many teams of astronomers have looked. Our present instruments could only detect brown dwarfs in the Sun's immediate neighbourhood, so instead of searching the whole sky, most astronomers have tried to check whether stars known to lie near the Sun have a brown dwarf companion.

Some of their early claimed successes have turned out to be false alarms: one 'brown dwarf' is probably a cloud of dust warmed by a neighbouring star, whereas another was merely a reflection in the infrared camera. Even where astronomers have identified a genuine heat-emitting companion, it is never quite clear if it is a genuine brown dwarf or simply a very cool red dwarf. For example, the red dwarf Gliese 569 and the white dwarf GD 165 both have companions with surfaces at temperatures near 2000 degrees Celsius: this is right on the borderline between the predicted temperature of the coolest red dwarf and that of the hottest brown dwarf.

Another method is more indirect: to look for the motion of a star as a brown dwarf companion swings it around their mutual centre of gravity. This technique is widely used in

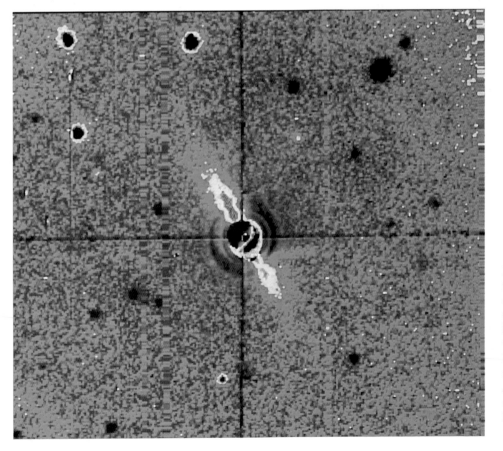

Is this a nearby solar system? The star beta Pictoris – 50 light years away – is surrounded by a disc of warm, dusty material that was discovered by IRAS. By blocking out the bright image of the star, Brad Smith and Rich Terrile were able to reveal the disc, seen here edge-on. Extending 60 billion kilometres from the star, the disc is larger than our Solar System. Although there is no trace of planets, the dust disc probably contains the raw materials – carbonaceous grains, silicates and ices – which will form into planets in the fullness of time.

hunting for planets, and one of the suspected planetary companions – to the star HD 114762 – may be a brown dwarf. But other planet-hunters have reported a complete lack of companions in the right range of mass to be a brown dwarf (see Box 4).

Recent searches within the Pleiades star cluster have yielded a good haul of potential brown dwarfs – now being checked out in detail. This star cluster is so young that the brown dwarfs have had little chance to cool down, and so they show up – albeit very faintly – to telescopes observing at red and infrared wavelengths.

In fact, some of the most convincing evidence is for brown dwarfs lying much futher away, in the Galaxy's halo (see Chapter 3). These brown dwarfs can account for an embarrassing shortfall of matter in the disc of the Milky Way. We can work out the average gravitational pull of the Galaxy's disc by meas-

---

BOX 4. Planets of other stars

It is only in our own neck of the galactic wood that we can hope to answer one question that intrigues astronomers and lay-people alike: do other stars have planets? Once we can make a census of planets around the nearest stars, we will have an idea of how common planetary systems are in the Galaxy at large – with the corollary that some of these planets, like the Earth, may be abodes of life.

The main problem is that planets are small and have no light of their own; and they lie close to stars that are intrinsically brilliant. So looking for planets near other stars is like trying to spot moths flying around a streetlight – at a distance of many kilometres.

In fact, no ordinary telescope on Earth can hope to see such planets directly, mainly because the Earth's atmosphere scatters the light from the bright star to such an extent that it would drown the planets' faint glow. The orbiting Hubble Space Telescope may just have a chance in the mid-1990s when astronauts fit a set of small mirrors to correct the defective vision of its mis-shapen main mirror.

A better way to search for planets is to look with an infrared telescope. At these longer wavelengths, the contrast between the star and its planets is much reduced, both because the star is dimmer in the infrared and because the planet glows with its own heat radiation in addition to the radiation it reflects from the star. At optical wavelengths, a planet would appear less than one-billionth as bright as its star, while at infrared wavelengths it could be all of one-millionth as bright.

Even so, the best infrared detectors at the moment are not capable of picking up an orbiting planet, although they can pick up the radiation from a 'brown dwarf', an object that is more massive than a planet but not as heavy as a star (see main text).

Ironically, it is easier for astronomers to pick out material around a star if it is scattered thinly rather than clumped into planets: the same amount of matter then has a much larger surface area to radiate infrared or to scatter starlight. In 1983, the Infrared Astronomical Satellite detected heat radiation from large discs of dust surrounding several nearby stars, including first-magnitude Vega and Fomalhaut. When America astronomers Rich Terrile and Brad Smith looked closely at one of these stars – beta Pictoris – with a ground-based optical telescope, they managed to obtain an image of the surrounding disc, considerably larger than the Solar System. Although we are seeing dust here, astronomers are confident that some of this matter must have condensed into planets.

A really large orbiting infrared telescope – with a mirror around 16 metres across and twenty-first-century infrared detectors – could easily detect individual planets around other stars. It could also analyse their radiation to tell if any of the planets have atmospheres containing oxygen – a hallmark of life. This highly reactive gas quickly disappears from a planet's atmosphere unless it is continuously regenerated: the Earth's air is unique in the Solar System because plant life produces oxygen from carbon dioxide.

At present, the most effective ways of searching out planets are indirect. As a planet orbits its parent star, the two in fact swing around their common centre of gravity. By carefully studying the motion of the star, astronomers can hope to deduce whether or not it is accompanied by planets.

The most venerable version of this 'search-by-wobble'

method is to study how a star moves across the sky: a lone star must travel in a straight line, but a star with a planet will follow a wavy path. Peter van de Kamp at the Sproul Observatory in Pennsylvania has spent decades in studying many nearby stars, and suspects that several – including Barnard's Star, the second closest system to the Sun – have at least one planet. But other astronomers have failed to confirm van de Kamp's results, and think that the 'wobbles' are not in the stars but in the Sproul telescope.

The best hope of detecting planets in this way will again come from telescopes in space. The guide system on the Hubble Space Telescope can – as a by-product of its main function – pin down the positions of stars with very high accuracy. And the Hipparcos satellite, designed to measure star positions and motions with high precision (see Chapter 3), will also be able to pick out wiggles in their paths. NASA scientists are planning a new generation of space astrometric telescopes that would definitely show wobbles of the nearest stars if any possess planets as massive as the gas giants of the Solar System.

There is another kind of 'wobble' that at the moment is showing more promise. As a star is swung round the centre of gravity by its lightweight companion, it moves alternately towards the Earth and away again. As a result, the Doppler effect moves its spectral lines first towards the blue end of the spectrum, then towards the red, in a cycle that repeats with each orbit of the planet.

Each spectral line moves by only a minuscule amount, and astronomers require a highly stable spectrograph to measure the shifts. Canadian astronomer Bruce Campbell has built just such a system – filled with highly poisonous hydrogen fluoride gas to provide a precise reference wavelength – and has observed many of the nearby stars with the large Canada–France–Hawaii Telescope on the Big Island of Hawaii.

In 1988, Campbell announced that nine of the 18 stars on his programme showed shifts in their spectra, indicating that they had planets with masses between 0.7 and 10 Jupiters.

Somewhat to his surprise, Campbell did not find any stars with a companion in the range 10 and 100 Jupiters – the 'brown dwarfs' that many astronomers believe should be extremely common in the Galaxy.

Some relatively conspicuous stars are on Campbell's list of planetary systems. They include Zavijava (beta Virginis), Alshain (beta Aquilae), Er Rai (gamma Cephei) and epsilon Eridani, which is a star fairly similar to the Sun in mass and spectral type. This research suggests that about half of all stars have planets as heavy as Jupiter.

The strongest evidence to date for planets has come from a most unexpected direction – from radio astronomers studying the motion of 'dead' stars called pulsars. In 1992, Aleksander Wolszczan and Dale Frail found that the pulses of radio waves from an obscure pulsar called PSR 1257+12 were not coming at regular intervals. Over the course of several months, the intervals became longer and then shorter again.

Like the Doppler shift in Campbell's measurement, the most likely cause of these changes was a planet moving around the pulsar and pulling it backwards and forwards. To their surprise, Wolszczan and Frail found they could fit the data perfectly with two planets. Each is three times heavier than the Earth, and they orbit the pulsar with 'years' that are about two and three Earth-months.

No-one expected to find planets in such a situation. A pulsar is left over when a star explodes as a supernova, and this cataclysm should destroy any planets that originally circled the star. Most likely, the pulsar's planets formed from debris left close to the pulsar after the explosion.

But this in itself is encouraging. If planets can be born in such a strange environment, then the formation of planets around ordinary stars must surely be a matter of course. In a few years, our improving techniques will show convincingly how common planets are. In the meantime, all the portents are encouraging – a good augury for the existence of an abundance of Earth-like worlds in our Galaxy.

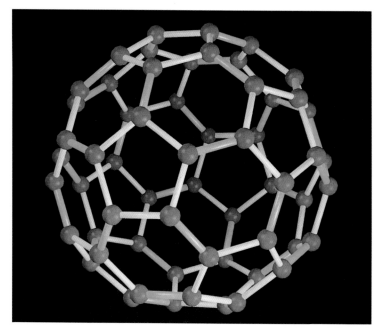

Extremely complicated molecules like buckminsterfullerene are expected to form around old stars like CW Leonis before they shed their envelopes as planetary nebulae. Cool, dying carbon stars are surrounded by 'soot' condensed from the outer layers of the red giant's atmosphere. The carbon atoms go through complex chemical reactions, building up enormous molecules in the process. Buckminsterfullerene consists of 60 carbon atoms (grey balls) linked by double bonds (red) and single bonds (white) in a near-spherical shape like a geodesic dome.

uring how far the stars stray away from the galactic plane before they are pulled back again. The figure turns out to be about twice the amount of matter that we know exists in the plane – in the form of stars and interstellar matter, but principally as the highly abundant red dwarf stars.

American theorist John Bahcall calculates that the 'missing material' could well be in the form of brown dwarfs. This would require brown dwarfs to be so common that the nearest one would probably be even closer to us than Proxima Centauri – just two or three light years away – and would be our closest neighbour on the interstellar scene.

Some astronomers have suggested that we may not even have to look that far to find the nearest brown dwarf. Although the smaller rocky planets in our Solar System undoubtedly built up from small fragments, there are good reasons for believing that the giant planet Jupiter – which contains two-thirds of the mass of the planetary system – may have condensed from a swirl of gas, in much the same way as a star forms. In addition, Jupiter is a strong source of infrared radiation: it gives out heat radiation as it gradually contracts.

The nearest brown dwarf, then, may not be an invisible object just at the limit of the most powerful infrared telescopes. It may be a familiar object in our night skies, shining more brightly than anything bar the Moon and Venus as it reflects the Sun's light. According to these theories, when we look at the planet Jupiter we are looking at the most common kind of object in the Milky Way.

CHAPTER 7

The Sagittarius Arm: within the Sun's orbit

A

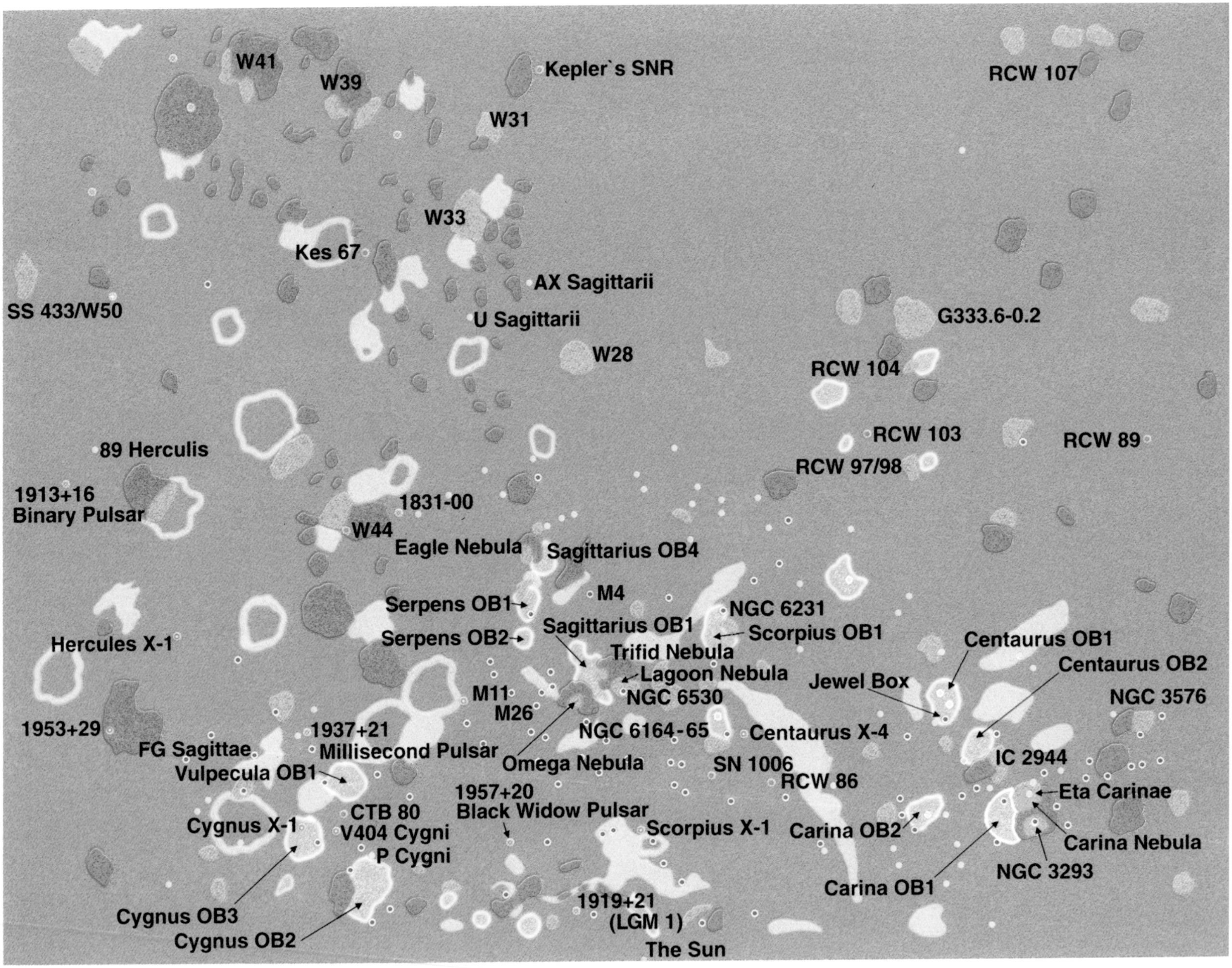

W41
W39
Kepler`s SNR
RCW 107
W31
W33
Kes 67
AX Sagittarii
SS 433/W50
U Sagittarii
G333.6-0.2
W28
RCW 104
89 Herculis
RCW 103
RCW 89
RCW 97/98
1913+16
Binary Pulsar
1831-00
W44
Eagle Nebula
Sagittarius OB4
M4
Serpens OB1→
NGC 6231
Sagittarius OB1
Scorpius OB1
Serpens OB2→
Trifid Nebula
Centaurus OB1
Hercules X-1
Centaurus OB2
←Lagoon Nebula
Jewel Box
NGC 3576
M11
NGC 6530
1953+29
M26
NGC 6164-65
Centaurus X-4
IC 2944
1937+21
FG Sagittae
SN 1006
Millisecond Pulsar
Omega Nebula
RCW 86
Vulpecula OB1
Eta Carinae
CTB 80
1957+20
Cygnus X-1
V404 Cygni
Black Widow Pulsar
Scorpius X-1
Carina OB2
P Cygni
Carina Nebula
Cygnus OB3
NGC 3293
Cygnus OB2
1919+21
(LGM 1)
Carina OB1
The Sun

1,000 Light Years

B

KEY TO MAP A OVERLEAF

KEY TO MAP A

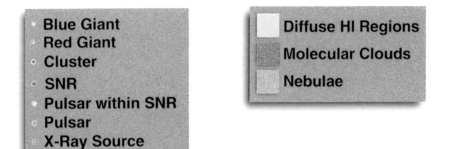

- Blue Giant
- Red Giant
- Cluster
- SNR
- Pulsar within SNR
- Pulsar
- X-Ray Source

- Diffuse HI Regions
- Molecular Clouds
- Nebulae

MAP A Great nebulae and dense molecular clouds mark the Sagittarius Arm as it winds past us on its way around the Galaxy. Famous names include the Eagle, Omega, Trifid and Lagoon Nebulae – making up one giant star-birth region – and the Carina Nebula further along.

This chart covers the further reaches of the Local Arm (lower left) winding inwards. It also maps what little is known of the still largely uncharted inner end of the Scutum–Crux Arm (upper right). This region is veiled from optical telescopes by heavy dust clouds, and it is difficult to untangle the longer-wavelength emission from different molecular clouds and nebulae.

MAP B The Sagittarius Arm is one of the Milky Way's two 'grand design' arms, stretching through more than one revolution around the Galaxy. But on the small scale, it breaks up into many individual features. The number of molecular clouds increases dramatically towards the central regions of the Galaxy, to culminate in a 'molecular ring' (top left) that lies near the end of the Galaxy's central bar.

As we travel in from the Sun's position towards the centre of our Galaxy, the scene becomes more crowded, busier and jammed with exotic and exciting objects. The next spiral arm in, the Sagittarius Arm, bristles with giant dark clouds, brilliant nebulae and bizarre dead stars – perhaps not surprising, as it is one of the Galaxy's two major spiral arms, a great continuous whirl of matter that reaches from the hub of the Galaxy to wind around for a full turn. In contrast, our local Orion Arm and the Perseus Arm further out are mere 'shingles' in our patchwork Galaxy.

Despite the promises of the Sagittarius Arm, the view from planet Earth is frustrating. Glance at a photograph of Sagittarius and Scorpius and you see dark swathes of dust blotting out much of the view. Some of these clouds lie in our own spiral arm, but the dense clouds of the Sagittarius Arm itself conspire to block our view of this arm and of objects that lie beyond.

By turning to radio astronomy and infrared telescopes, we can penetrate the dust. But this clearer view in some ways adds to the confusion. Objects at different distances all lie on top of one another, so that the foreground denizens of the Sagittarius Arm are confused with a background of remote stars and nebulae, stretching all the way through to the far side of the Galaxy. And there is an extra problem with measuring distances. If we cannot pick out individual stars in a nebula, astronomers rely on measuring the speed imposed on it by the rotation of the Galaxy (see Box 3 of Chapter 3). When we look inwards from the Sun's position, however, any particular velocity can correspond to two very different distances. So, for example, one particular faint radio source at the borders of Scorpius and Ara is either a fairly insignificant nebula at a distance of 10 000 light years or one of the most massive nebulae in our Galaxy, lying 50 000 light years away on the very far fringes of the Milky Way system. In this chapter, then, we will keep to the well-explored paths that lead from the Sun no more than halfway to the Galaxy's centre.

In fact, let us start right at home. The map that accompanies this chapter shows us the inner parts of our own arm, the Orion Arm, as it heads off into Cygnus. Let's follow the trail of this arm first. We will pick up on a couple of objects that in fact appear on the map in Chapter 5, but which we have left until now because they provide a unifying theme to the inner part of the Orion Arm: it is a rich hunting ground for

some of the most unusual specimens of one of the Galaxy's most exotic creatures, the neutron stars.

The story begins in 1967, when radio astronomers in Cambridge were investigating the way in which some radio sources 'twinkle' as their radiation travels through the wind of gas that blows incessantly outwards from the Sun. One of the researchers, Jocelyn Bell, noticed that the paper chart from the recorder occasionally showed a burst of radiation that looked different from the twinkling that she was expecting. With her supervisor, Tony Hewish, Bell rigged up a faster chart recorder, and found that each patch of 'scruff' on the chart in fact consisted of a set of equally spaced pulses, repeating with uncanny regularity in a period of just over a second.

Bell and Hewish jokingly named this radio source 'LGM1' – for 'Little Green Men'. They had little doubt that it was really a natural kind of radio transmitter, and this suspicion was confirmed when they found three more pulsating radio sources: it was too much to believe that four sets of little green men were trying to contact us simultaneously! The Cambridge astronomers dubbed these sources 'pulsars', and a theorist, Tommy Gold, soon realized that they are very compact stars – neutron stars – that are spinning around once every second or so, and sweeping a beam of radiation across us to produce pulses – like the beam of a rotating light-house lantern. The discovery of a pulsar in the Crab Nebula showed that a rotating neutron star was the core of a star that had exploded as a supernova (see Chapter 4).

The first pulsar to be discovered lay in the constellation Lacerta, comparatively near the Sun. Astronomers soon dropped its early nickname LGM1, and it now goes only under its catalogue number, PSR 1919+21. Apart from its proximity to the Sun, which made it easy to detect, this is a very typical pulsar. Astronomers have now found over 400, with periods that range from little more than a thousandth of a second to several seconds. The total in our Galaxy must be far larger than we have catalogued, both because it is difficult to detect pulsars that lie more than 10 000 light years away and because we only pick up pulsars whose beams sweep over us. For every pulsar with a beam we can detect, there are probably four or five that are tilted at the wrong angle. Our Milky Way, astronomers estimate, must contain about half a million pulsars altogether.

A pulsar's radio emission is generated as fast electrons spin

The view towards Sagittarius and Scorpius is blotted out by huge swathes of interstellar dust cutting across the star clouds. Although radio waves and infrared radiation penetrate the dust, the picture can sometimes become even more confusing, as we then see lots of objects lined up on top of each other along the distant line of sight. This photograph of the Scorpius−Sagittarius region was take in April 1986, when Halley's Comet was superimposed against the star clouds of the Milky Way.

This view towards the centre of the Galaxy, made with the infrared satellite IRAS, penetrates much of the intervening dust. As well as showing clearly the plane of the Milky Way, it reveals several of the nebulae that belong to the Sagittarius Arm (blue blobs in this image). The Trifid/Lagoon complex is just to the left of the galactic centre; the bright nebula at far left is NGC 6604.

Twenty-one years on from their discovery of pulsars, Jocelyn Bell and Tony Hewish stand in the bleak East Anglian landscape amongst the wires of the radio telescope that discovered them – the four-acre array. Since 1967, over 400 pulsars have been found, some spinning round in little more than a thousandth of a second.

around in its strong magnetic field. The Crab Pulsar, for example, has a magnetic field about a million million times more powerful than the Earth's. But this magnetism dies out when the pulsar is around ten million years old – just a fraction of the age of the Galaxy. As a result, the Milky Way must contain around a thousand 'dead' pulsars – undetectable neutron stars – for every one that is alive and shining.

Not far from PSR 1919+21, however, there is a prime example of a dead neutron star that has been reactivated. This object, Scorpius X-1, was discovered in 1962, when American astronomers launched a Geiger counter on a rocket, ostensibly to search for X-rays from the Moon. Instead, they found a powerful source of X-rays in the constellation Scorpius. It was the first cosmic X-ray source to be discovered beyond the Solar System. Much further research has shown that this source – and many others in our Galaxy – consists of a neutron star in

orbit around an ordinary star. Gas from the normal star is falling towards the neutron star, and swirling around it in an extremely hot disc that generates copious amounts of X-rays.

Further along the inner edge of the Orion Arm, behind the dark clouds of Vulpecula and Cygnus, we come across one of the most unusual of the known pulsars. The prosaically catalogued PSR 1957+20 shot to fame in 1988, when its bloodthirsty habits earned it the nickname the Black Widow Pulsar. Orbiting this rapidly spinning neutron star is something much lighter in weight: although the pulsar is slightly more massive than our Sun, the companion weighs in at only one-fiftieth of a solar mass. We happen to see this system edge-on, so the companion star regularly blocks off the radiation from the pulsar. From the length of this eclipse, we can work out the size of the lightweight star – and it is relatively vast, at least 50 per cent bigger than the Sun. The companion star's weak gravity should not be able to hold onto such inflated outer layers, and the neutron star's gravity should be ripping its gases away.

Astronomers have therefore concluded that the secondary in the PSR 1957+20 system is not a stable star at all. It is a cloud of gases boiling away from a small embedded star that is being heated by the pulsar's energy. Originally, the companion was an ordinary star rather lighter in weight than the Sun. But the energy from the fast-spinning pulsar has heated up its outer layers, expanding them and then ripping them off into space. This exposes the inner layers of the companion, which then expand and evaporate in turn. Long-exposure images of this system show a wake of gas from the evaporating star, left behind as the two stars waltz along through space in their dance of death. The 'black widow' tactics of the pulsar will eventually destroy its companion altogether.

As we travel on along this inner part of the Orion Arm, we are heading almost directly away from the Earth, in the direction of the constellation Cygnus. Beyond the local dark clouds, we find a couple of giant regions where stars have been born recently. Some 5900 light years from the Sun lies a loose

FACING PAGE A glowing cloud of gas surrounds the bizarre 'Black Widow Pulsar' (PSR 1957+20) and its unfortunate companion. The cloud appears to be matter boiling off a small star that is being heated by the pulsar's energy. Eventually, all the star's matter will evaporate into space, and the pulsar will have destroyed its companion.

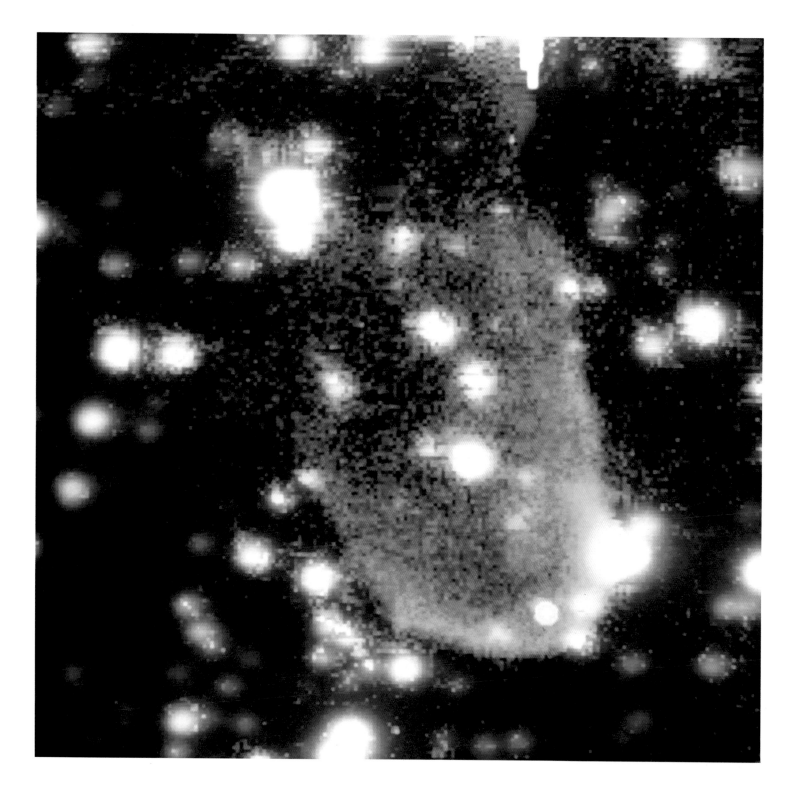

grouping of young stars called the Cygnus OB2 Association. Its most massive member, too faint to appear in any major star catalogue, is one of the most luminous stars in the Galaxy. 'VI Cygni No. 12' is a million times brighter than the Sun. Despite its great distance, this star would shine in our skies as a star of the first magnitude, if it were not obscured by dense clouds of dust that block off 99.99 per cent of its light.

Nearby is another massive star – and this time one that has made its mark in our skies. P Cygni is 500 000 times brighter than the Sun and is normally just visible to the naked eye. In August 1600, however, it leapt to ten times its normal brightness. P Cygni is clearly an unstable star, and present-day spectra show that it has thrown off three separate shells of gas, at speeds of several hundred kilometres per second.

An X-ray view of a black hole – or more accurately, of the energetic disc of gas swirling round a black hole. This is an image from the Rosat X-ray telescope of Cygnus X-1, the first suspected black hole in our Galaxy. Since its discovery in 1971, more and more evidence has accumulated that Cygnus X-1 consists of an ordinary star orbiting around a black hole that is at least six times heavier than the Sun.

Next comes another rapidly spinning pulsar, which was literally pointed out to astronomers. An early radio survey at Caltech turned up a supernova remnant called CTB 80. It has an extraordinary arrow shape that points, it later transpired, to the pulsar that was born when the supernova exploded. Astronomers believe that the pulsar was given a kick in the course of the explosion, and it has travelled sideways until it has run into the shell of gases ejected by the supernova and distorted it.

Like other supernova remnants, CTB 80 is a source of X-rays. But it is far outclassed by its near neighbour, Cygnus X-1. Here matter is falling from a large hot star onto a compact companion that is probably not a neutron star, but is almost certainly a black hole – the first to be detected in our Galaxy. Nearby is another double star, V404 Cygni. It also contains a dark compact object, which is indubitably a black hole.

Beyond CTB 80 and Cygnus X-1, we come across the fastest-spinning pulsar currently known. PSR 1937+21 is more commonly called the Millisecond Pulsar, because its period of spin is little more than a thousandth of a second (a millisecond). It created a sensation in astronomical circles when it was discovered in 1982.

Until then, the record for rapid spinning belonged to the Crab Pulsar, which turns once in 33 milliseconds (30 times per second). This in itself seemed extraordinarily fast for a star that was more massive than the Sun. The Crab Pulsar is almost certainly the youngest neutron star we know, as it is the core of a star that blew up only a few centuries ago. The strong magnetic field of a young pulsar acts as a brake to slow down its rotation, so it rotates at a more and more leisurely pace as it grows older. Because the Crab Nebula is so young, astronomers thought that even a new-born pulsar would spin little more quickly than the Crab Pulsar.

In 1982, however, researchers from Princeton were checking out a suspicious radio source in the constellation Sagitta, using the world's most sensitive radio telescope at Arecibo in Puerto Rico. To their amazement, the equipment showed pulses coming in at a rate of one every 1.56 milliseconds: they were observing a star that was spinning round 642 times in one second!

As the months and years went by, astronomers found that this pulsar is keeping up its rapid pulse-rate pretty well: it is slowing down, but at a very gradual rate. Evidentally, its

BOX 1. **Black holes in Cygnus**

On the seventh anniversary of Kenya's independence day in 1970, a rocket blasted off from a converted oil platform off the country's coast. It carried aloft the first satellite designed to scan the sky for sources of X-rays in the Universe. Because of its launch date, the American satellite was christened 'Uhuru' – the Swahili word for freedom – and it was indeed to liberate astronomers' minds from the straitjacket of thinking that all objects in the Universe can be seen by some means or other. Uhuru was to discover the ultimate in cosmic concealment: the first black hole.

Physicists had thought of the concept of black holes almost two centuries earlier: a body could exist with gravity so strong that light cannot escape. In the 1910s, Albert Einstein and his colleagues refined the idea. According to Einstein's theory of relativity, nothing can travel faster than the speed of light. An object that traps light will thus prevent anything else from escaping. Things can fall into this dark region, but nothing can ever escape again – a concept that led American physicist John Wheeler to coin the graphic phrase 'the black hole'.

By its very definition, a black hole should be invisible. Yet its effects may be detectable – as Uhuru discovered in 1971. The satellite pinpointed the position of an X-ray source in Cygnus, called Cygnus X-1. By then astronomers suspected that most X-ray sources in our Galaxy consisted of a compact neutron star that was in orbit around a normal star. Gas ripped from the companion falls towards the neutron star, to form a blindingly hot disc that generates X-rays.

To begin with, Cygnus X-1 seemed to fit the same picture. Optical astronomers located a normal star, HDE 226868, at the same position as the X-ray source. This star lies 7500 light years from the Sun, and its spectrum shows that it is orbiting a small unseen companion, once every six days.

But a detailed investigation started to show something odd about the system. The star's spectrum indicates that it is a fairly massive star, around 20 to 30 times heavier than the Sun. The orbital characteristics showed that the unseen star was one-third to one-half the weight of the primary – in other words, it was at least as massive as six Suns. There are good theoretical reasons for believing that there is an upper limit to the mass of a neutron star, of about three solar masses. Only one kind of object can be as small as a neutron star, but weigh as much as the unseen companion in Cygnus X-1 – a black hole.

Astronomers are a notoriously conservative breed, however. At first, many were reluctant to include such an odd object as a black hole in the roll-call of the Galaxy. They argued that the compact object might consist of two neutron stars in a close orbit, or that the main star in the system might be a lightweight star close to the Sun that was mimicking the appearance of distant massive star. But all further research has only strengthened the original case for a black hole existing in the Cygnus X-1 system.

Sceptical astronomers had argued that if one dying star could collapse to become a black hole, in Cygnus X-1, then there must be others – and where were they? Over the years, researchers have turned up several more examples like Cygnus X-1, including an X-ray source in our neighbouring galaxy, the Large Magellanic Cloud, that makes an even stronger case. In the LMC X-3 system, the visible star weighs as much as six Suns. The spectrum also shows that it is being swung around by an invisible companion that is even more massive – well over the weight limit for neutron stars.

But the clinching evidence for black holes in our Galaxy has come, ironically enough, from a double star system that is a near neighbour of Cygnus X-1. Astronomers in the 1930s had seen V404 Cygni light up as a nova; when the star erupted again in 1989, researchers using the Japanese satellite Ginga picked up a flood of X-rays, too. This high-energy radiation meant that V404 Cygni did not contain a white dwarf, like ordinary novae, but must harbour a compact object with a powerful gravitational pull – either a neutron star or a black hole.

In 1991, optical astronomers turned the powerful William Herschel Telescope – normally reserved for distant quasars – onto this star. The star's speed showed that it is orbiting a dark companion that weighs in at something between 8 and 15 Suns. Even the sceptics are hard pushed to explain away this massive, dark and compact object as anything but a black hole.

magnetic field is too weak have a serious effect on its rate of spin. This weak magnetic field, and the lack of any supernova remnant around the Millisecond Pulsar, suggest that it is comparatively old. So why is it spinning so fast?

The answer, as in so many good detective stories, seems to lie in the accessory who has left the scene of the crime. Astronomers think that this neutron star once had an ordinary star in orbit around it. When the neutron star was already very old, and its magnetic field had decayed to virtually nothing, the companion star began to dump gas onto it – as we find in Scorpius X-1 today. The swirling gas imparted a sideways force to the neutron star that increased its rate of rotation, 'spinning it up' like a Victorian child with a whip and top. The fate of this couple was eventually to break up. The companion star may have exploded and flung the neutron star off into space; or the rapidly spinning pulsar may have acquired enough energy to destroy its erstwhile companion, as the Black Widow Pulsar is doing today.

Whatever the past history of the Millisecond Pulsar, this distant beacon may serve us a practical purpose on the Earth. With little magnetic field to slow it down, and no companion to disturb its motion, this neutron star is almost the physicist's 'ideal flywheel on frictionless bearings'. Allowing for the star's very gradual spin-down, we can use its regular rotation as the most perfect clock. Measurements over the past decade have confirmed that the pulsar keeps time at least as well as the best clocks on Earth. Atomic clocks drift gradually, because of external effects like tiny changes in temperature. If we want a standard of time-keeping that we can rely on, decade after decade, we might be best to slave our terrestrial clocks to the natural time-keeper at the Cygnus end of the Galaxy's Orion Arm.

Near the Millisecond Pulsar is a star that has been called the 'Rosetta Stone' of stellar evolution. FG Sagittae is a variable star that changes irregularly in brightness and temperature, and has thrown out a shell of gas that looks like a planetary nebula half a light year across. The star itself is unique because its spectrum reveals traces of many unusual elements in its outer layers, including samarium, yttrium and zirconium. Astrophysicists have long believed that elements heavier than iron can be made in one of two ways: 'rapidly', in a supernova explosion, or 'slowly' in the outer layers of an old star. The elements found in FG Sagittae are just those that we would expect to be cooked up slowly within a star, so it provides both strong support for the basic theory and a chance to sample the ingredients of the stellar cauldron.

FG Sagittae is dredging up these interior gases, it seems, because it has reached an unstable period right at the end of its life. Astronomers believe that this star has already been through the red giant stage; it has even ejected some of its matter as a planetary nebula, and in the last century it appeared as a superhot star of the kind that we find in the centre of other planetary nebulae. Since then, FG Sagittae has experienced a sudden hiccup of fresh nuclear reactions. Some of the reaction products have risen to the surface of the star, while the sudden burst of energy has made FG Sagittae swell up and cool to the temperature of the Sun.

At the far end of the Cygnus stretch of the Orion Arm, 11 000 light years away, we find another X-ray binary, Hercules X-1. Here some of the gas falling onto a neutron star is channeled by the compact star's magnetic field to its poles, so the X-rays are beamed outwards like the radio waves from a conventional pulsar. As a result, the X-rays from Hercules X-1 pulse regularly, with a period just over a second, as the neutron star rotates.

Beyond Hercules X-1, we cross from the end of our spiral arm into part of the great Sagittarius Arm. Optical astronomers have trouble in probing this part of the Sagittarius Arm, because it is largely obscured by the dust clouds in the foreground. But radio waves have no such problem, and they have led astronomers to another exotic pulsar.

The Binary Pulsar – as its name suggests – was the first radio-emitting pulsar to be found in orbit with another star. Although astronomers have now discovered several more 'binary pulsars', this one (catalogued as PSR 1913+16) is still the most important, for it provides our most accurate test of General Relativity, Einstein's theory of gravity.

The Binary Pulsar lies in the outer part of the thick Sagittarius Arm; on the inner side of the same region of the arm is one of the Galaxy's real oddities: SS 433 – described in one newspaper headline as 'The Star that is Coming and Going at the Same Time'! SS 433 lies some 18 000 light years away from us, in the constellation Aquila. Its 'name' reflects its inclusion in a catalogue of stars that show strong emission lines, drawn up by American astronomers Bruce Stephenson and Nicholas Sanduleak in the 1960s. This star lies in the middle of a large

BOX 2. **A galactic laboratory for Einstein**

In 1974, Joe Taylor and Russell Hulse of the University of Massachusetts at Amherst were using the big Arecibo radio telescope to search for more new pulsars. Among the 40 new sources, Hulse found one with a rapid pulse rate. It repeated once every 59 milliseconds, and was then the fastest known after the Crab Pulsar. More surprisingly, the pulses were not exactly constant but changed slightly over a period of 7.75 hours. The most natural explanation was that the pulsar was orbiting another, unseen, star. As the pulsar comes towards us, the pulses are crowded together; they are stretched out as the pulsar moves away in the other half of the orbit.

All we can measure for this system is the string of radio pulses from one of the stars, but Taylor and his colleagues have been able to analyse these signals to provide basic data about the system. They find that both stars are 1.4 times as heavy as the Sun. This is the normal mass for a neutron star, so the unseen companion is almost certainly a neutron star that is not beaming radio waves. The two stars also follow distinctly oval-shaped paths around one another.

Taylor quickly realized that this system was an answer to a physicist's dream: it provided a laboratory for testing Albert Einstein's theory of gravity, General Relativity. When gravity is weak, General Relativity gives almost the same results as Newton's venerable law of gravity but Einstein's theory makes some very different predictions when gravity is strong. Unfortunately, the Earth's gravity is so weak that we cannot easily compare the two theories in the laboratory. Over the past few decades, astronomers have been able to test some of Einstein's predictions using the gravitational pull of the Sun: for example, it should deflect and slow down the path of light and radio waves from spacecraft, stars and quasars lying behind. But these effects are comparatively small, and the results controversial.

In the Binary Pulsar system, however, strong gravity is everywhere. If Einstein is right, then the orbits of the stars should be changing markedly from what would be predicted by Newton's theory. The beauty of the system is that we not only find strong gravitational forces here, but can use the regular pulses to provide a probe of the effects of this gravity.

As a result, Taylor's team has been able to confirm that Newton's theory is hopelessly inadequate at explaining the Binary Pulsar's characteristics, while Einstein's theory fits with almost uncanny accuracy. Conversely, they have been able to use some of the 'relativistic' effects on the string of pulses to derive an orbit that is now more accurately known than any other orbit outside the Solar System – even though the Binary Pulsar is 15 000 light years away! For this work, Taylor and Hulse were awarded the 1993 Nobel Prize for physics.

Among the effects that defy Newton's theory is the fact that the two stars do not repeat exactly the same orbit time after time; the orbits swing around in space, at just the rate predicted by Einstein's equations. Even more remarkable, the two stars are gradually spiralling together. Einstein's theory insists that the mutual orbit must shrink, because the stars lose energy by emitting gravitational radiation – just as an accelerating electric charge broadcasts electromagnetic radiation. On top of that, the rate at which the stars lose orbital energy is exactly equal to the rate at which the gravitational waves should be carrying off energy. Although they have yet to pick up gravitational radiation directly, most astronomers take this evidence from the Binary Pulsar as proof that such radiation exists.

radio-emitting cloud, W50, that the Dutch radio astronomer Gart Westerhout had catalogued in the previous decade.

In the late 1970s, several researchers independently found themselves studying SS 433. Canadian radio astronomers were making a routine check for radio emission from SS stars, and put SS 433 down as a positive detection, without realizing it was in the centre of the extended radio source W50. Simultaneously, radio astronomers in Cambridge and Australia were looking for supernova remnants with a neutron star in the centre, and found a point-like source of radiation within W50 – without realizing it was the already catalogued star SS 433.

The latter observations prompted British optical astronomers Paul Murdin and David Clark to look within W50 using the Anglo-Australian Telescope. They expected that any star remaining from the supernova explosion would be very faint, and at the limit of even this large telescope. To begin with, they

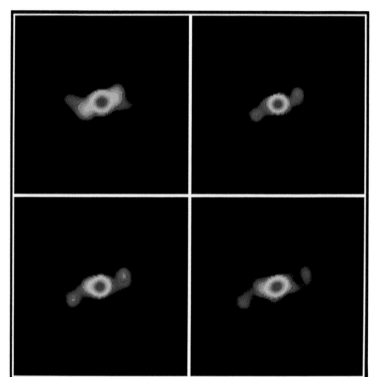

So far unique in the Galaxy, SS 433 appears to be a normal star in orbit about a very dense companion – probably a neutron star. The neutron star snatches gas from the normal star, which forms an accretion disc around it. What makes the system so unusual, however, is that the disc somehow emits jets of gas at a quarter of the speed of light – a cosmic record. These 'time lapse' radio images of SS 433, obtained over four months with the Very Large Array telescope in New Mexico, show the jets emerging from the system. Like spray from an eccentric lawn-sprinkler, the jets swing round, revealing that the disc is tilted.

wavelengths, marching up and down the spectrum in a regular rhythm that repeated every five and a half months. The average position of the lines was just that of hydrogen, the commonest element, and the changes were undoubtedly caused by a changing Doppler Effect as we viewed clouds of hydrogen moving alternatively towards us and away again. There is one simple way to explain the regular changes in the lines: the star is squirting out two jets of gas, in opposite directions, and as the jets swing round, like a garden lawn sprinkler, the jets point alternately towards and away from us.

What was more remarkable than this alternating motion was the speed involved. The gas is travelling at a quarter the speed of light. Although we know places in the Universe where sub-atomic particles, like electrons, are moving at a velocity very close to the speed of light, nowhere have we met ordinary gas moving so fast. Some extraordinary source of energy must be powering these jets of gas.

One clue came from studying the spectrum of the star SS 433, neglecting the strange moving lines. The light comes from a fairly ordinary giant star, but one that is in orbit about an unseen companion, probably a neutron star. Once again, we have a system where gas from an ordinary star is falling towards a neutron star, and presumably forms into a hot disc around the smaller star. In the case of SS 433, however, the disc is not content to produce just X-rays. In some way, it manages to channel a small amount of the infalling gas into two jets, which shoot off at incredible speed from either side of the disc. The disc is slightly tilted and the gravitational pull of the large star makes it gradually wobble around over a period of five and a half months, swinging the jets around.

Radio telescopes have now revealed the wobbling jets, stretching out to one-fifth of a light year to either side of SS 433 itself. The view in X-rays shows gas drifting even further out, extending the line of the jets to a hundred light years or more. Where the gas from the jets hits the inside of the supernova remnant's shell of gases, it pushes out 'ears' to either side of W50. This discovery has led some astronomers to think that the large radio source W50 is not an ordinary supernova remnant at all, but a large bubble of gas inflated by the jets from SS 433. This could explain why it is so large – with a maximum width of 600 light years, W50 is much larger than any other supernova remnant. Even at its great distance, it is four times the width of the Full Moon in the sky.

intentionally avoided the brightest star in the area, thinking it must be a star near the Sun that happened to lie in front of W50. But, after checking out a couple of fainter stars, they looked at the bright star – and found it had a very unusual spectrum: they had 'rediscovered' SS 433. Their detailed spectra, however, revealed some odd spectral lines that Stephenson and Sanduleak had not seen.

American astronomer Bruce Margon had access to several telescopes, and he was able to watch the spectrum from week to week. To his amazement, the unusual lines changed their

So far, SS 433 is unique among the many strange objects in the Galaxy. Certainly there cannot be many more in the Milky Way, as this example draws attention to itself even though it lies 18 000 light years away.

From SS 433, let us move up along the Sagittarius Arm, clockwise round the Galaxy so that we come towards the regions of the arm that lie between the Sun and the centre of the Milky Way. Near a supernova remnant called W44 – another name from Westerhout's catalogue – we find a giant group of dense molecular clouds, containing as much matter as two million Suns. Oddly enough, at least one star must have been born, lived and died to create the remnant W44, but there is little sign of any other stars forming here. Near this complex there is another binary pulsar.

As we reach the region of the Sagittarius Arm that is nearest to the Sun, the scene becomes more exciting. This major arm of the Galaxy begins to live up to its promise, as we come across a series of great regions of star-birth, among the largest we know in the Galaxy. As well as brilliant young stars, we find nebulae containing over a thousand solar masses of hot gas and immense molecular clouds – reservoirs massive enough to form millions of stars each. Despite some veiling by foreground dust in space, this region of the Sagittarius Arm is well-placed for observation from the Earth.

The first of these massive regions of star-birth lies in the direction of the constellations Sagittarius, Scutum, Scorpius and Ara. A huge but patchy molecular cloud has largely turned into stars within the past ten million years. Where stars are still being born, however, we are treated to some of the most beautiful nebulae in the sky: the Eagle, the Omega, the Trifid and the Lagoon.

The first we come across on our journey up the arm is M16, the Eagle Nebula. Long-exposure photographs show a glowing cloud of gas that is roughly shamrock-shaped, with the poised silhouette of an eagle in the centre. The eagle is in fact only one of several 'elephant trunk' structures in the nebula – each an elongated region of dust that has been protected from evaporation by a particularly dense clump at one end.

The 'eagle' nickname is comparatively recent, and dates only from the introduction of photography. In a small telescope, what stands out is the cluster of bright stars that has formed here. These stars were born six million years ago, and they have largely dispersed the gas immediately surrounding them:

in addition, the active young stars have blown a bubble of hydrogen to the side of the nebula some 200 light years across.

Appearing as a neighbour on the sky, but slightly closer to us according to the most recent distance estimates, is the Omega Nebula (M17). In contrast to the Eagle, the view through a telescope shows intricate loops of glowing gas with no noticeable cluster of stars. The total brightness of most nebulae is augmented by the combined light from the newly born cluster of stars within it. The Omega Nebula lacks this extra boost to the visual observer, and in terms purely of the brightness of shining gas, it is one of the most brilliant nebulae to be seen from the Earth.

The French astronomer Philippe de Chéseaux discovered the Omega Nebula in 1746. He commented 'it has a shape quite different from the others, it has the perfect form of a ray or the tail of a comet, its sides are exactly parallel'. A larger telescope shows a prominent curve springing from one end of the bar, giving rise to perhaps the most appropriate – as well as beautiful – nickname, the Swan Nebula. Larger apertures still showed the curve bending right round on itself, like the loop on a capital Greek omega.

Modern instruments have shown that the Omega Nebula is exceptionally bright when viewed at infrared wavelengths, too – a sure sign that star-birth is still continuing. Maps made with infrared and radio telescopes have indeed revealed that the Omega Nebula marks just the first stage in what will be a prolonged sequence of star-birth. The curve of the Omega (or the head of the Swan) marks the boundary between the tenuous glowing gases of the nebula and a dark dense molecular cloud. The cloud extends 250 light years from the Omega Nebula and contains enough matter to make a million stars.

The nebula itself shows where star-birth has begun at one end of this long gas cloud. The newly formed stars are mainly hidden from our view inside the molecular cloud, but their light shines out sideways to illuminate the gases in the Omega Nebula, which forms a 'blister' of hot gas on the side of the cloud. The Orion Nebula is a blister that we observe end-on and so can see the hot stars that lie in a recess in the dark molecular cloud. To observe the hidden stars that illuminate the Omega Nebula, astronomers have recently used infrared cameras, which pay scant regard to the dust in the way.

The radiation pressure from these young stars will compress the gas that lies further along the molecular cloud, and start

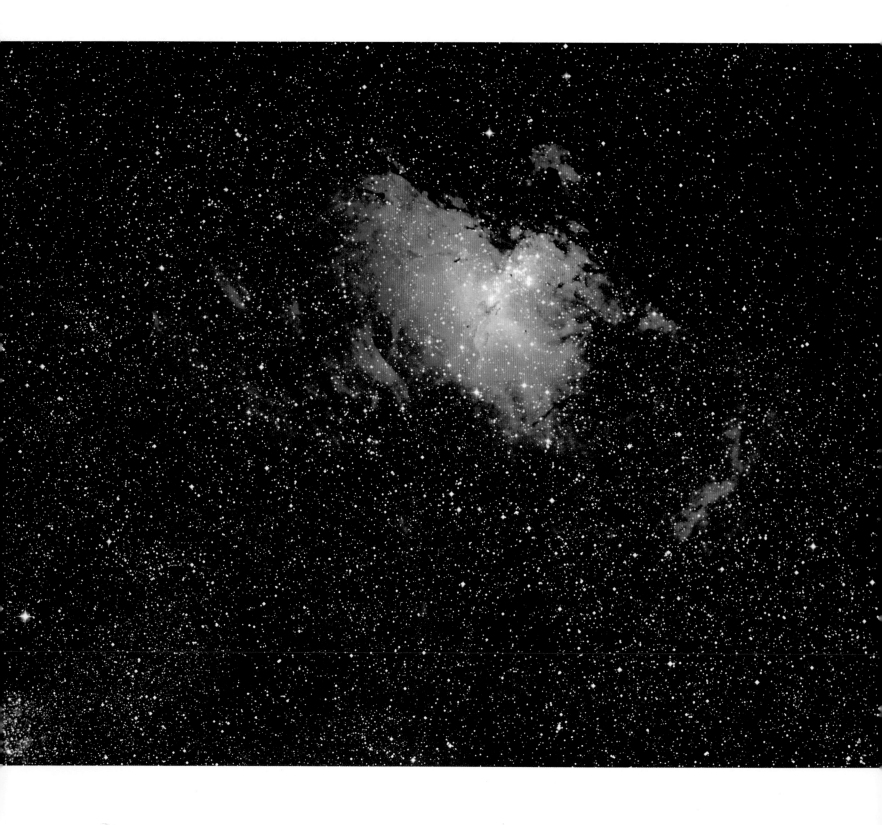

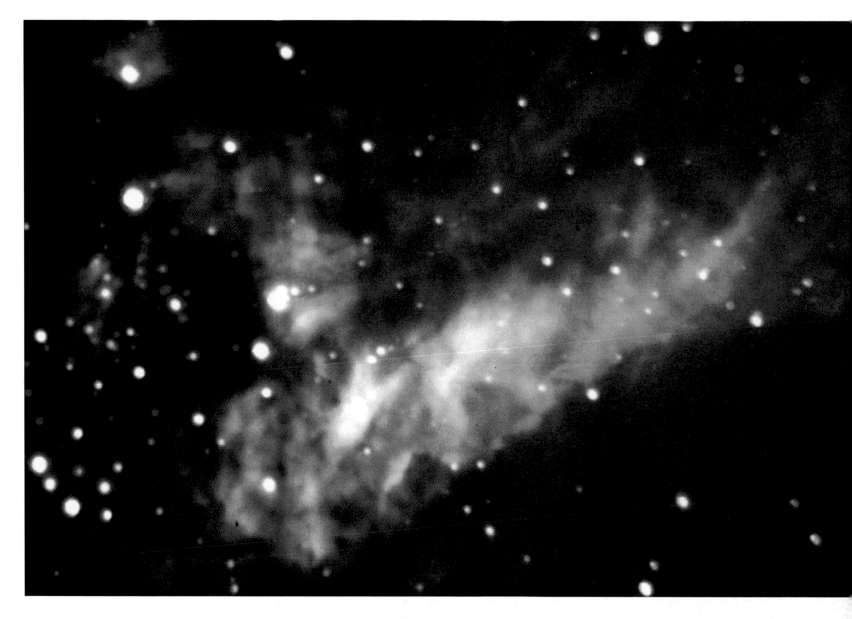

FACING PAGE The Eagle Nebula gets its nickname from the dusty silhouette of an eagle hovering in front of the glowing gas at the top of the nebula. The nebula surrounds a cluster of young stars born six million years ago, which are the most prominent feature of the region seen in a small telescope (the 'eagle' is only revealed on long-exposure photographs). These active young stars have blown a bubble of hydrogen to the side of the nebula some 200 light years across.

The Omega Nebula is a neighbour of the Eagle Nebula, but is very slightly closer to us. This image brings out the shape that has given it its alternative name: the Swan Nebula. Star-formation is taking place in the 'head' of the swan, and the swan's curved neck marks the boundary between the glowing gas of the nebula and the dense molecular cloud. The cloud contains enough matter to form a million stars. Some young stars, which appear exceptionally red, can be seen near the swan's neck – partly because they are intrinsically red, and partly because they are reddened by interstellar dust.

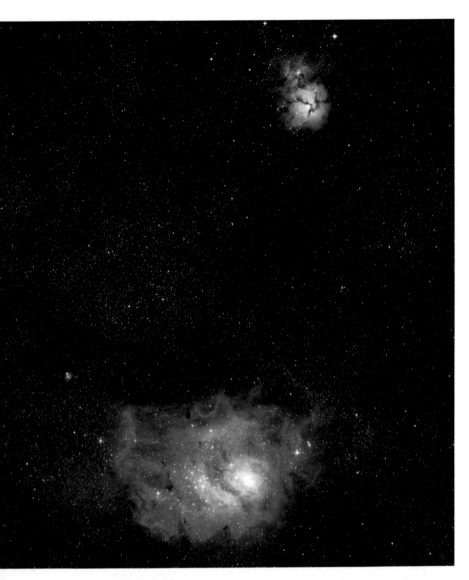

Over 5000 light years away, the Trifid (top) and Lagoon Nebulae are still among the most striking in the sky. The Trifid gets its name from the prominent dust lanes that divide it into three; the Lagoon from the dark patch that crosses in front of its glowing gases. The Lagoon envelops a cluster of stars (NGC 6530) whose members are so massive and bright that they can be seen with the unaided eye. These stars are a mere two million years old; but younger still is 9 Sagittarii, which also lies in the Lagoon. It is among the half-dozen most luminous stars in the Galaxy, shining brighter than a million suns. At 100 solar masses, it is one of the heaviest stars, and it is also one of the hottest, with a surface temperature approaching 50 000 degrees Celsius.

these regions on the path to star-birth. Over the next few million years we can expect a trail of star-formation to spread along the currently existing molecular cloud until the whole lot is converted into stars.

Next in line come a rather dissimilar pair of nebulae that really are close neighbours. The Trifid Nebula, M20, is not named after the alien monsters of a science fiction novel. It is simply Latin for 'split into three parts'. The English astronomer Sir John Herschel – son of Sir William – described it thus: 'singularly trifid, consisting of three bright and irregularly formed nebulous masses . . . at their interior edges they enclose and surround a sort of three-forked rift or vacant area, abruptly and uncouthly crooked and quite vacant of nebulous light'.

Despite its name, the Trifid is not really divided into three parts. It is a sphere of glowing gases, and the dark divisions are bands of dust that obscure some of its light. The spherical shape of the nebula does not mean that this is a perfectly formed ball of gas in space – the cloud of gas undoubtedly extends beyond the nebula itself. The nebula's shape is a result of the way it is illuminated. The gases are lit up by the ultraviolet radiation from a very compact cluster of stars at its centre. This radiation is absorbed as it makes the hydrogen in the nebula glow, and so at a certain distance – roughly the same in all directions – the ultraviolet just runs out. As a result, the illuminated region is fairly accurately spherical. The physics of this process were first worked out by the Danish astronomer Bengt Strömgren, and the Trifid Nebula is the most easily observed example of a 'Strömgren Sphere'.

Much larger and brighter is the Lagoon Nebula, M8, named for the dark lagoon-like patch that runs across the middle of its glowing gases. The Lagoon is a classic example of continuing star-birth. It contains a young cluster of stars that is now largely free from gases; newly hatched stars just emerging from the cocoon; and protostars still hidden in the dense molecular cloud behind the nebula itself. The Lagoon is a 'blister' that we look squarely into, and wide-field photographs show that its surrounding halo of dark clouds blocks virtually all the light from more distant stars.

The cluster of stars in the Lagoon Nebula (NGC 6530) is only two million years old. It contains stars so massive and bright that, despite its distance of 5200 light years, the cluster is visible to the naked eye. It was one of the earliest star clusters to be recorded, and was noted by England's first Astronomer Royal, John Flamsteed, around 1680 as the 'nebulosity preceding the bow' of Sagittarius.

When newly born, the stars of NGC 6530 must have provided the radiation that made the Lagoon Nebula shine. But its stars have now been overtaken in vigour by a new generation, in particular by a single brilliant star called 9 Sagittarii that lies on the other side of the dark dust 'lagoon'. Its catalogue-boring name belies the fact that 9 Sagittarii is among the half dozen most luminous stars we know in the Galaxy, shining more brightly than a million Suns. It is also one of the hottest ordinary stars, with a temperature of 50 000 degrees Celsius; and one of the most massive, weighing in at 100 Suns. Even though it is over 5000 light years away, we can pick it out with the naked eye, and its number shows its place in Flamsteed's catalogue of naked-eye stars.

Even the brilliant 9 Sagittarii is about to be overtaken in importance by the newest couple of stars to be born from the dust behind the nebula. The brightest part of the Lagoon Nebula is lit up by a newly emerged star, spotted by John Herschel and called Herschel 36. Next to it is the Hourglass Nebula. This seems to consist of gases flowing from the poles of a very young star, which itself is obscured by a band of dust around its equator – dust that may well coagulate into planets. Astronomers have had a stroke of luck in having the Hourglass laid bare like this. Infrared and radio telescopes have suggested that most stars go through a stage like this when they are still cocooned within the dark molecular cloud and are hidden from optical telescopes. In the Lagoon Nebula, the nearby hot star Herschel 36 has stripped away the surrounding dark gases and exposed the Hourglass to our view.

The end of this region of frantic star-birth is marked by a rather pretty cluster of stars in Scorpius, NGC 6231. This cluster is visible to the naked eye, and through a telescope forms a sparkling setting of brilliant white and bluish-white gems. It formed well before many of the other clusters in this starbirth region, but even so its age of five million years makes it exceptionally young on the galactic scale.

Here we also find signs of star mortality. Optical photographs show an attractive S-shaped nebulae, NGC 6164-65, on the borders of Ara and Norma. After years of debate, astronomers now believe this gas marks the funereal pall of a star that lies at its centre, and will shortly explode as a super-

nova. It will follow in the path of two other stars in the vicinity that have 'gone supernova' in the past few hundred years, and now exist only as the supernova remnants, RCW 86 and SN 1006. Comparatively few of the supernova remnants in our Galaxy have come from supernovae that were recorded on Earth – most of the explosions were either too distant or too far back in the past – but the supernovae that produced both of these remnants were probably recorded in the historical annals of China or Europe.

Also in this region is Centaurus X-4, a neutron star 'starcorpse' dragging matter from a companion star. It leapt to the attention of astronomers – and others – in 1969, when Centaurus X-4 produced a burst of X-rays powerful enough to trigger the American Vela satellites. These were not astronomical telescopes, but satellites designed to watch for X-rays from nuclear weapon tests. Centaurus X-4 was the first 'X-ray nova' to be discovered. In an ordinary nova explosion, a blob of gas emits a burst of light as it falls onto a white dwarf star. If a similar blob falls onto a neutron star, it reaches a much higher energy, and therefore temperature, and the force of the impact generates X-rays instead.

Further on, we come across a second huge region of star-birth in the Sagittarius arm. Whereas the region we have just passed included several fairly large nebulae – the Eagle, the Omega and the Lagoon – the Carina region of star-birth is dominated by one enormous glowing cloud of gas. In the great Carina Nebula, 9000 light years from us, some of the brightest and most massive stars we know are being born right now.

Star-birth started in the Carina region 10 million years ago, and some of the first-born stars now form young clusters in the outskirts of the region, around and behind the Carina Nebula as seen from the Earth. In total, the Carina star-formation region stretches 5000 light years along the spiral arm. As a result it covers a wide range of sky, from the Southern Cross (Crux) through parts of Centaurus into Carina itself.

On this journey, we enter the Carina region by one of its most spectacular sights: the Jewel Box. This beautiful little star cluster lies close to the Southern Cross and the Coal Sack in the sky, but in reality it is totally unconnected. Whereas the Coal Sack and the two brightest stars in the Cross lie about 400 light years away, the Jewel Box is almost 20 times further

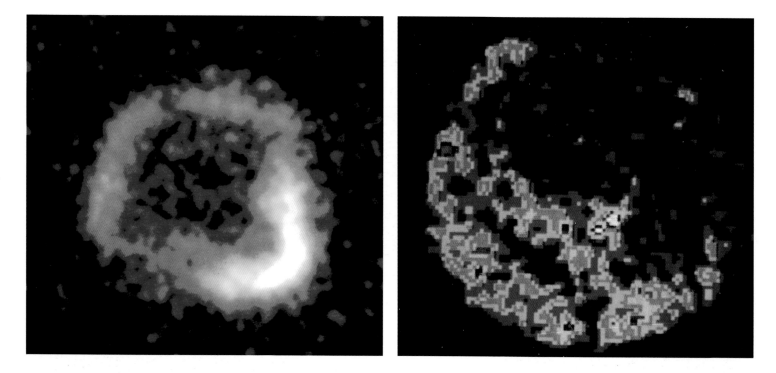

All that remains of the earliest recorded supernova is a distorted shell of hot gas known as RCW 86. The exploding star was observed by Chinese astronomers in AD 185. Its remnant was first detected almost 1800 years later by radio astronomers, and its unusual shape appears clearly in this X-ray image from Rosat. The star probably exploded asymmetrically, with fast-flying ejecta causing the bright bulge in the outline of the remnant.

Exosat X-ray image of the wreck of the most brilliant supernova ever seen, the exploding star of 1006. Reported by watchers in the East, and even by monks in Switzerland, the supernova shone a hundred times brighter than Venus – brilliant enough to cast shadows. Today, we can detect X-rays from the expanding remnant, which originate in hot gas at a temperature of about 10 million degrees Celsius. On this colour-coded image, the most intense X-rays come from the white regions, the least intense from those coded blue.

FACING PAGE The shell-shaped nebula NGC 6164-65 surrounds a peculiar hot star (HD 1448937) some 25 000 times brighter than the Sun. The nebula, which is about seven light years across, is surrounded by a cavity enclosed by a circular shell of ionized hydrogen and dust. Until recently, astronomers were uncertain whether the whole complex – which is about 130 light years across – was young or old. Now they believe that HD 1448937 is a relatively evolved, massive star that has ejected a substantial amount of material. An earlier age estimate of 200 000 years – derived from the motions of the nebulosity – is thought to correspond to a time when radiation from the star broke through a thick dust shell, and set off the evolution of the various nebulae.

OVERLEAF Divided by dust lanes into the shape of an exotic flower, the Carina Nebula appears twice as large as the Full Moon – yet it is 9000 light years away. The nebula, which is 20 times larger than the Orion Nebula, has spawned a small number of exceptionally massive stars. One is the most luminous star in the Galaxy, HD 93129A. Four million times brighter than the Sun, this star weighs in at 120 solar masses and has a surface temperature of 52 000 degrees Celsius. Because it emits most of its radiation at ultraviolet wavelengths, it is not at all obvious in a conventional telescope.

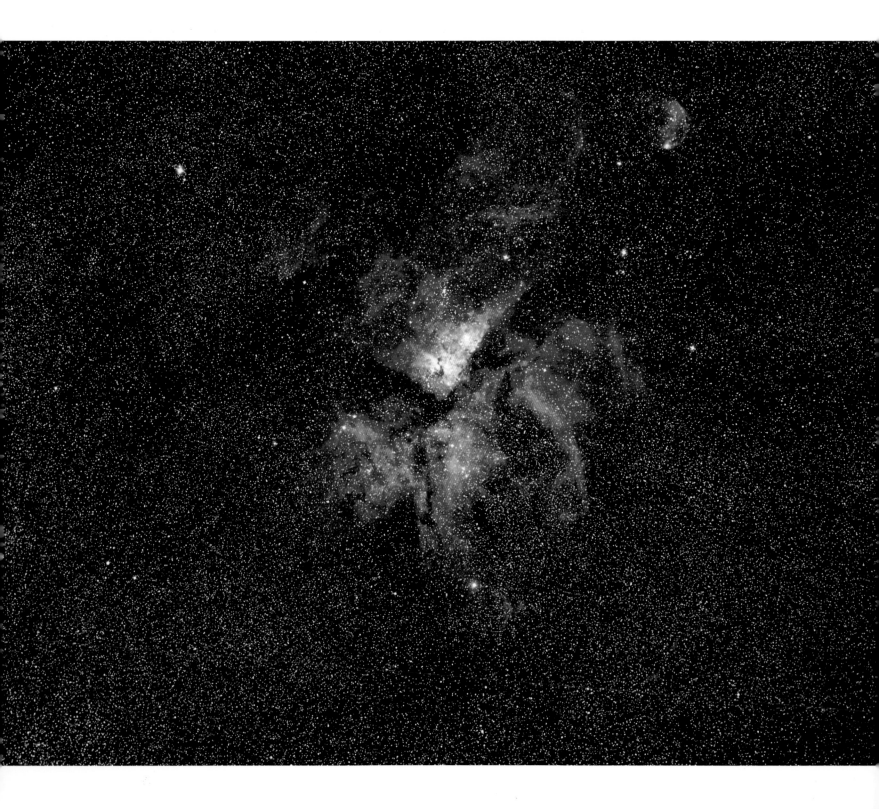

BOX 3. **Supernovae at the southern gates**

The supernovae seen from the Earth in 185 and 1006 occurred in the same part of the Galaxy – on the outer edge of the Sagittarius Arm, the region of this mighty arm that is closest to the Sun. They are also both record-setters. The supernova of 185 is the earliest that we can locate in the historical records, while the supernova of 1006 was the brightest that our ancestors ever noted.

In December 185, the official astronomers in China wrote 'a guest star appeared . . . it was half as large as a mat; it displayed the five colours and it scintillated. According to the standard prognostication, this means insurrection . . . Wu-kuang attacked and killed Ho-miao, the general of chariots and cavalry, and several thousand people were killed'.

The British astronomical historian Richard Stephenson and astronomer David Clark have attempted to track down the remains of this exploded star. The Chinese recorded that the star appeared in their constellation of Nan-mên, the Southern Gate. Stephenson has checked old Chinese star maps, and has found that Nan-mên consists of the two bright stars alpha and beta Centauri, which rose just a few degrees above the southern horizon as seen from Lo-yang, in central China, the capital of the ruling Han dynasty.

Between alpha and beta Centauri, radio astronomers have found four sources of radiation with the characteristics of supernova remnants. Three are faint and probably distant – almost certainly the remains of supernovae that exploded thousands of years ago. But one source seems comparatively close to the Sun, and is emitting radio waves so powerfully that it must have resulted from a comparatively recent supernova explosion. This remnant, RCW 86, also has some glowing strips of nebulosity along its edge.

According to measurements of its X-ray emission, from the Exosat orbiting observatory, RCW 86 lies 4500 light years from the Sun, on the outer edge of the Sagittarius Arm. This makes it one of the closest of the supernovae recorded in historical times, and so it should have been one of the brightest.

Chinese astronomers were notoriously unconcerned with recording the brightness of stars, but they did carry on observing the 185 supernova with the naked eye for one and a half years. With our modern knowledge of how these stars fade, Clark estimates that the star must have been around magnitude −8 at maximum – about 20 times more brilliant than Venus at its most magnificent.

But even this brilliance was surpassed by the supernova of 1006. This celestial light-show was recorded not only in China, Japan and Korea but also in Arabic and western countries that had little interest in variable stars at that time. Observing from Egypt, Ali ibn Ridwan noted that 'this spectacle appeared in the zodiacal sign Scorpio . . . The sky was shining because of its light. The intensity of its light was a little more than a quarter of that of moonlight'. From this – and other descriptions – Stephenson and Clark have calculated that the star reached a peak magnitude of −9.5. It was then a hundred times brighter than Venus at its most brilliant, and its light must have been intense enough to cast shadows.

The most remarkable observation was from Switzerland. The monks at the monastery of St Gallen saw 'a new star of unusual size . . . causing alarm. It was seen in the inmost limits of the south, beyond all the constellations which are seen in the sky'. Stephenson interprets the account as meaning the star only just rose above the great mountains of the Alpstein to the south of St Gallen. With this information, and other data from the far eastern astronomers, there is only one radio source that could be the remnant of this supernova. It lies about 3500 light years from us, once again on the near side of the great Sagittarius Arm.

Radio and X-ray telescopes show that this remnant (usually known as 'SN 1006') has a strikingly symmetrical and near circular shape; photographs reveal faint and delicate filaments moving outwards at just the rate we would expect for an explosion that occurred about a millennium ago.

away. If we brought it as close to us as the Coal Sack, the Jewel Box would shine as brightly as Jupiter.

The Jewel Box is seven million years old, and its brightest star, kappa Crucis, has evolved into a red giant. Its ruddy colour forms a vivid contrast to the other blue and white stars in the cluster. 'The stars . . . have the effect of a casket of variously coloured precious stones', noted Sir John Herschel, and his description led to the cluster's nickname.

More spread out – and therefore much less spectacular in a small telescope – is another star cluster of the same age, which lies in the Sagittarius Arm behind the nearby star lambda Centauri. But long-exposure photographs show that it is surrounded by a remarkable gaseous nebula, IC 2944. The hot stars are dispersing the gas into space, to create an impression of folded red velvet curtains, and suspended in front are several irregularly shaped black specks. Discovered by the South African astronomer David Thackeray in the 1950s, 'Thackeray's globules' were originally dense patches in the dark cloud that condensed into stars. Other dense blobs coagulated into stars, but the Thackeray globules never made it to stardom. Instead, they are now being destroyed by the radiation from the young stars that did form.

Surrounded by a retinue of handsome young star clusters, such as the 'alternative jewel box' (NGC 3293), the great Carina Nebula itself almost beggars description. Although 9000 light years away, the nebula is easily visible to the naked

Seven million years old, the stars of the Jewel Box cluster lie close to the Southern Cross and the Coal Sack in the sky. However, the Jewel Box is 20 times further away, and would appear as bright as Jupiter if it were as close as the Coal Sack. One of the stars in the cluster, kappa Crucis, has already evolved into a red giant. Its ruddy colour makes a striking contrast with the blue of the other members of the cluster, which have yet to evolve.

FACING PAGE Hanging eerily in front of the glowing gases of IC 2944 are dense lumps of dark cloud known as Thackeray's Globules. Discovered by David Thackeray in the 1950s, these are the final shreds of the dark cloud that gave birth to the young stars of the cluster IC 2948, seen in the background here. The ultraviolet radiation from the stars is nibbling away at the globules, and will ultimately destroy them.

eye and appears twice as large as the full Moon: in actual size, it is 20 times wider than the widely vaunted Orion Nebula. Through binoculars you can see that the nebula is divided into sections by dark dust lanes, whereas a small telescope shows the region is 'beautiful beyond description', in the words of leading Australian amateur astronomer Ernst Hartung. In the nineteenth century, Sir John Herschel felt impelled to write: 'Nor is it easy for language to convey a full impression of the beauty and sublimity of the spectacle . . . ushered in as it is by so glorious and innumerable a procession of stars, to which it forms a sort of climax'.

The nebula is a huge 'blister' where stars have begun to form, over the past two million years, in the front face of a giant molecular cloud that extends to several times its own size and contains enough matter to make a million stars. Here the forces of star-birth are gathered into a single giant nebula, rather than several medium-sized nebulae. As a law of Nature, more massive stars are progressively rarer, so we are likely to find the heaviest stars in regions where a lot of matter has condensed into a very large number of stars. This makes the Carina Nebula a hotbed of heavyweight stars. And because more massive stars are generally brighter, we also discover in this nebula some immensely bright stars, including the two most luminous stars known in the Galaxy.

According to a definitive survey by the Dutch astronomer Cornelis de Jager, the record holder is a star known only by its catalogue number, HD 93129A. It is about four million times brighter than the Sun, and weighs in at 120 solar masses. With a surface temperature of 52 000 degrees Celsius, HD 93129A is one of the hottest ordinary stars. As a result, it emits most of its radiation at ultraviolet wavelengths and is not as prominent as we might expect in ordinary telescopes.

At the heart of the Carina Nebula is the other contender for 'most luminous star', eta Carinae. Apart from its sheer luminosity, this star is one of the most enigmatic and unstable stars known. Its remarkable history began in 1677, when the young British astronomer Edmond Halley fled the tedium of Oxford University to make the first survey of the southern skies. Halley recorded this star as being fourth magnitude, but noted nothing else exceptional. Later visitors to the southern hemisphere, however, found that this star was sometimes several times brighter than Halley's estimate. In 1837, while Sir John Herschel was observing from South Africa, the great British astronomer saw eta Carinae flare up: six years later, it became the second brightest star in the sky, almost rivalling Sirius. Herschel was led to muse: 'What origin can we ascribe to these sudden flashes and relapses? What conclusions are we to draw as to the comfort or habitability of a system depending for its supply of light and heat on so uncertain a source?'

Knowing the distance to eta Carinae, we can now work out that in 1843 it was around five million times brighter than the Sun. After that outburst, its light faded to magnitude 6. But in the 1960s, the pioneers of infrared astronomy discovered that if we observe the sky at a wavelength around 20 microns, eta Carinae is now the brightest infrared source in the heavens. Including the infrared emission, the star's total luminosity is several million times more than the energy output of the Sun.

A small backyard telescope can give us a clue to what is going on. Under high magnification, eta Carinae does not look like a star at all: it is a small egg-shaped nebula with brighter patches that give it a humanoid shape – leading Thackeray to christen it 'the Homunculus' (little man). The Homunculus consists of dust that was ejected in the 1843 outburst and is now shrouding the star itself. According to American astronomer Kris Davidson, who specializes in investigating massive stars, the star hidden within the Homunculus is still five million times brighter than the Sun, as luminous as it was in 1843. Its light, however, cannot escape directly to space; instead it heats up the dust, which re-emits the energy as infrared radiation.

The spectrum of the Homunculus shows that the star inside is extremely unstable. It is spewing out gas from deep in its interior, gas containing a telltale amount of nitrogen that has been forged in its central nuclear fires. Almost certainly, eta

FACING PAGE Lying at the heart of the Carina Nebula, eta Carinae is not only one of the most massive and luminous stars known, but also one of the most unstable. This Hubble Space Telescope image shows the star cocooned in dust ejected during an outburst in 1843, when eta Carinae flared up to become the second-brightest star in the sky. Astronomers believe that eta Carina may be five million times brighter than the Sun. However, the star is not expected to last much longer. The gas it is currently spewing out includes nitrogen, which comes from deep in the star's core. This highly volatile star will almost certainly explode as a supernova within a few thousand years.

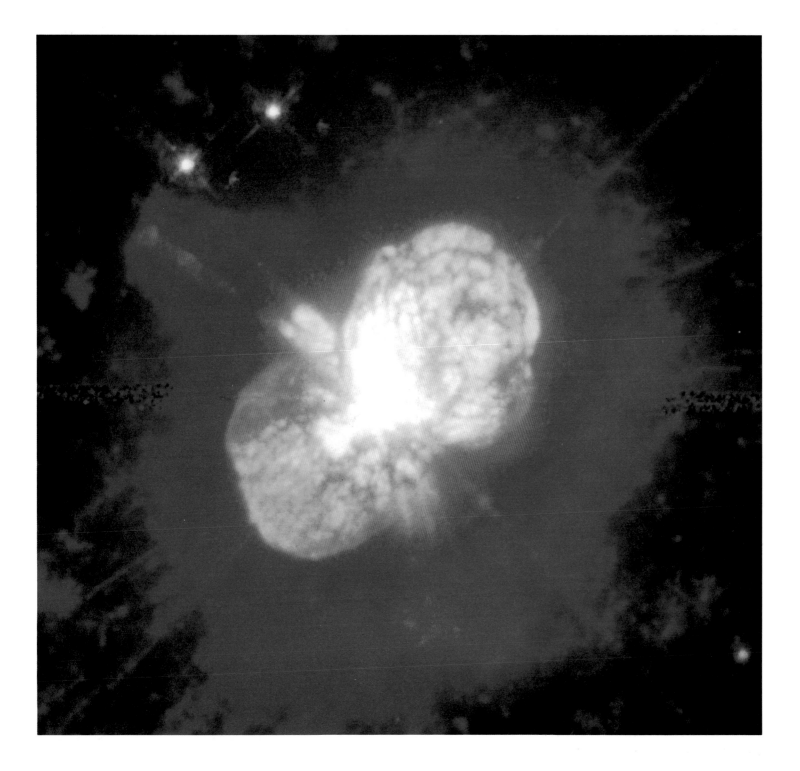

The 'twin' nebulae NGC 3576 (right) and NGC 3603 (left) lie close to the Carina Nebula and were once thought to be parts of a single nebula. But whereas NGC 3576 does form part of the Carina complex, NGC 3603 lies three times further away – as evidenced from its reddening by the foreground dust in NGC 3576. NGC 3603 is the most massive nebula in the Galaxy visible to optical telescopes, and contains twice as much gas as the Carina Nebula. It is lit by a dense cluster of extremely hot young stars, at least three of which are as luminous and massive as any of the stars in the Carina Nebula.

Carinae is nearly at the end of its life. Within a few thousand years – perhaps in only a few centuries – eta Carinae will explode as a supernova, outshining everything in the sky apart from the Sun and the Moon.

Close to the Carina Nebula is an apparently well-matched pair of nebulae, with the NGC numbers 3576 and 3603. For many years, astronomers regarded tham as parts of a single nebula. But appearances are deceptive. Here we have a prime example of how perspective and veiling by interstellar dust can confuse our picture of the Galaxy.

The picturesque loops of NGC 3576 do indeed form part of the Carina complex of star-formation and they make up a fairly run-of-the-mill nebula. But its apparent twin in the sky could not be more different. NGC 3603 lies far in the background, three times as distant from the Sun, and it is the most massive nebula in the Galaxy that is visible to optical telescopes. Although it contains well over 10 000 solar masses of glowing gas – more than twice as much as the Carina Nebula – NGC 3603 appears comparatively dim because it is obscured by the dust that lies around the foreground nebula NGC 3576.

Detailed studies have shown that NGC 3603 is lit up by a very tight cluster of extremely hot young stars, practically hidden from Earth by the intervening dust. At least three are as bright and massive as any star in the Carina Nebula, and these obscure stars are certainly contenders for the title of the Milky Way's most luminous star.

If we could see the Galaxy from the outside, NGC 3603 would outshine any nebula in the Perseus Arm, the Orion Arm or the portion of the Sagittarius Arm we have investigated in this chapter. We would also see that NGC 3603 is part of the Sagittarius Arm, even though it lies behind the Carina region of this arm as seen from the Earth. Here we are looking through the nearer part of the arm to the farther reaches of the Sagittarius Arm as it curves around the Galaxy.

Not far beyond NGC 3603, which lies 25 000 light years away, our view becomes so obscured by foreground dust that optical telescopes cannot hope to see even the most masssive nebulae. A few thousand light years further along the arm, for example, radio telescopes suggest that there is a nebula (known only by its coordinates, as G298.4−0.9) that is every bit as massive and luminous as the Carina Nebula, but is totally hidden at optical wavelengths.

The highly obscured nebula G333.6−02 – immortalized in verse by the Australian astronomer John Storey – shines brilliantly in the infrared, as revealed here in an image from the Anglo-Australian Telescope. It lies 15 000 light years away, and its total luminosity amounts to three million suns.

Dust also gets in our eyes when we try to look at what lies inside the Sagittarius arm. Again, radio telescopes (and, to a certain extent, infrared telescopes) have helped us to map the gas and dust in the next arm in. This 'Scutum–Crux Arm' is probably a patchy feature, consisting of separate shingles a few thousand light years in length – more like the Orion and Perseus Arms than the major Sagittarius Arm. Radio astronomers have traced parts of the Scutum–Crux Arm by locating its dense molecular clouds and hot nebulae, and have also picked out several supernova remnants where massive short-lived stars have exploded.

In the Scutum–Crux Arm, radio telescopes have revealed another huge glowing nebula, suffering the unmemorable catalogue number G333.6−0.2. According to the radio results, its total luminosity amounts to some three million Suns, although dust obscures all this light from our telescopes. Infrared astronomers have found evidence for the birth of energetic young stars here, shooting out powerful jets of gas that shock and excite the molecules in the surrounding cloud. It was only the second place where astronomers found shocked carbon monoxide – ensuring this nebula's place in verse (see Box 4).

Further in towards the Galaxy's centre, space becomes thicker and thicker with molecular clouds. Here there is an enormous reservoir of gas and dust, waiting its turn to be processed into stars. The problem of determining the Galaxy's structure here becomes truly daunting – what with the distance from the Sun, the obscuration and the piling up of all these clouds, one behind the other, as seen from our position in the galactic plane.

What information we do have suggests another spiral feature, the Norma Arm, beyond the Scutum–Crux Arm. It begins some 12 000 light years from the Sun, and about the same distance from the galactic centre. The radio observations of its molecular clouds and nebulae suggest that the Norma Arm is the beginning of the Galaxy's second major spiral arm, mirroring the Sagittarius Arm. The Norma Arm disappears behind the Galaxy's centre, and after a complete revolution, it comes around outside our local Orion Arm and the Perseus Arm, to end in a shower of fragments, the Outer Arm of the Galaxy.

BOX 4. **An ode to starbirth**

There is nothing more uninspiring in a book on astronomy than to find that some fascinating object languishes under a mere catalogue number. Australian researcher John Storey, however, took this as a challenge when he reported to a high-level astronomy conference on his observations of the highly-obscured nebula G333.6−0.2 in the constellation Norma. He was using the Kuiper Airborne Observatory – a telescope-carrying aircraft – to search for a spectral line from shocked carbon monoxide, previously found only in the Orion Nebula. To liven up the subject, he presented his findings in verse.

> The problem with this new-found source
> I hardly need to warn you,
> Is that it's too far south to see
> From sunny California.
>
> And so to us the Kuiper came;
> In May last year it made it.
> The cost was astronomical,
> But NASA mainly paid it.
>
> . . .
>
> The line is here, I kid you not,
> This dip here's just the sky.
> To see the peak you simply need
> A good impartial eye.
>
> Intensity is really weak
> It's two point nought by ten
> To the minus eighteenth power
> (In watts per square cm).
>
> That's thirty times as weak as we
> Detected in Orion.
> No wonder it took several years
> Of concentrated tryin'.
>
> But now we have not one, but two
> CO sources, it is true:
> Orion and this G333
> Point six, minus, nought point two.

CHAPTER 8

The centre of the Galaxy

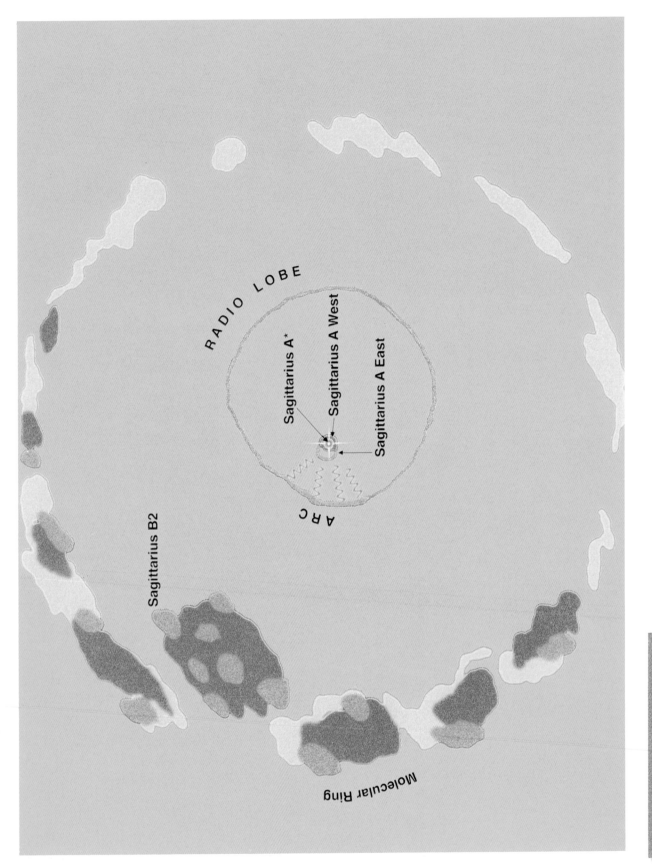

RADIO LOBE

Sagittarius A*

Sagittarius A West

Sagittarius A East

ARC

Sagittarius B2

Molecular Ring

100 Light Years

A

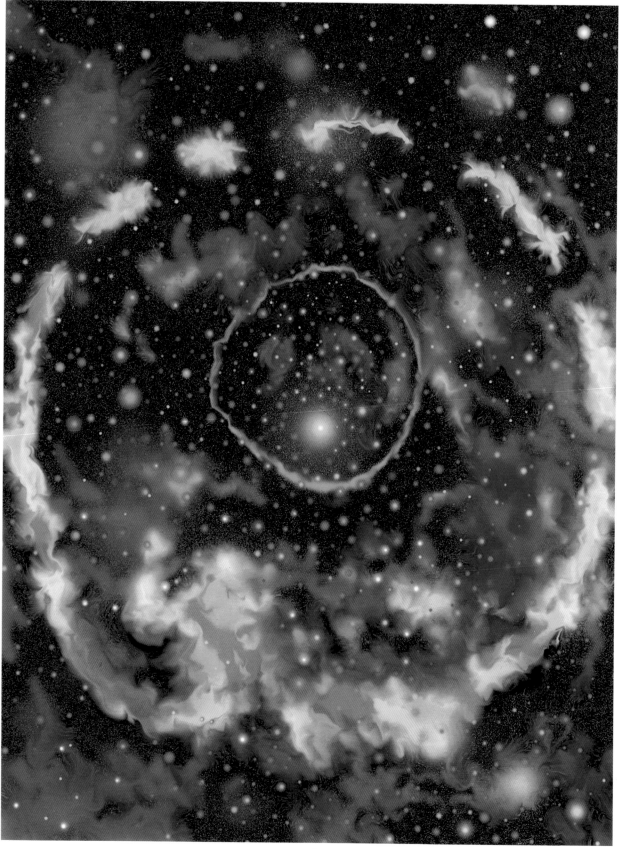

B

KEY TO MAP A OVERLEAF

KEY TO MAP A

MAP A A simplified chart of the complex region at the galactic centre shows the features that can be definitely located. Many other objects – X-ray sources, peculiar supernova remnants and moving gas clouds – have been found near the Galaxy's core, but their position along the line of sight ('up or down' on the page) is not yet known.

The magnetic influence of the nucleus is probably responsible for the Arc and the radio lobe (which stretches vertically in and out of the plane of the page). The Galaxy's heart is marked by a compact radio source, Sagittarius A*, surrounded by a small unique spiral-shaped nebula, Sagittarius A West.

MAP B Stars crowd closer and closer together towards the Galaxy's heart, which is surrounded by a ring of gas and dust clouds and scattered globular clusters. At the very core, the stars are held tightly in a compact cluster. A tiny source of energy appears in this bird's-eye view as a minuscule point of light – a miniature version of the brilliant quasars that are found in many other galaxies. The tight gravitational bondage and the tiny central powerhouse both suggest that the centre of our Galaxy harbours a black hole as massive as a million Suns.

'We have to face the fact that there is some outstanding peculiarity at the galactic centre. To radio astronomy belongs the merit of having discovered this peculiarity, whose study will, in our opinion, become one of the central problems of astrophysics and cosmology'. These words, written in 1958 by the Soviet astrophysicist Iosef Shklovskii, were indeed prophetic. He wrote them as astronomy was entering its current 'wavelength revolution', when astronomers were starting to get to grips with the baffling variety of unknown objects revealed through the new windows on the Universe. Since then, many of these new objects have become reasonably familiar – but not the galactic centre. Almost 40 years on, astronomers are still studying the 'outstanding peculiarity' – and they still aren't certain what it is.

One difficulty in probing the galactic centre is our suburban position within the Galaxy itself. Distance is not really the problem; it's more that our line of sight to 'downtown' goes straight through the most thickly populated regions of the galactic plane. This is also the 'slice' that happens to be most densely packed with interstellar dust. These grains only comprise about one per cent of the Galaxy's mass, but over several thousand light years they have a considerable effect. The net result is that whatever lies at the centre of the Galaxy is dimmed so much that only one photon of light in 100 billion gets through to us.

Fortunately, the grains are virtually transparent to longer-wavelength radiation. This means that most of the observations of the galactic centre are done at radio or infrared wavelengths, and – as Shklovskii pointed out – it was radio astronomy that provided the first indications of strange goings-on 'downtown'.

In the 1950s, Dutch and Australian radio astronomers collaborated on a major project to map the whole Galaxy at a wavelength of 21 centimetres (more fully described in Chapter 3). As well as revealing the cold hydrogen gas in the Galaxy's spiral arms, the survey let astronomers peer, for the first time, well inside the Sagittarius Arm. Most obvious were three strong radio sources close to the centre of the Galaxy, all in the general direction of Sagittarius. Sagittarius A – flanked by Sagittarius B and C – seemed to mark the exact centre of the Galaxy itself (although in those early days of radio astronomy, the resolution of details was poor).

But it was the area just outside the centre of the Galaxy that proved to be the most intriguing. The Dutch project leader, Jan Oort, noticed that whereas the outer spiral arms followed straightforward orbits about the galactic centre, those closest in did not. He first drew attention to an arm just inside the Sagittarius Arm – the 'Three Kiloparsec Arm' – which, as well as circling the galactic centre, appeared to be moving outwards with a velocity of 50 kilometres per second. On the far side of the Galaxy, he identified an arm racing away from us at 135

The Russian astrophysicist Iosef Shklovskii (1916–85) was the first to recognize that the radiation from the Crab Nebula was synchrotron radiation, which, up until then, had only been detected in man-made particle accelerators. He also realized – as long ago as the 1950s – that unique events were taking place at the centre of our Galaxy.

The galactic bulge shows up dramatically in this view of Scorpius and Sagittarius taken from Mount Graham, Arizona. The dust in the galactic plane masks our view of the galactic centre so effectively that only one visible photon in 100 billion gets through to us.

BOX 1. **The distance to the galactic centre**

The main problem in measuring the distance to the centre of the Milky Way has always been the clouds of dust that prevent us from seeing the Galaxy's downtown regions directly. In the early twentieth century, Harlow Shapley at Harvard took an intellectual 'giant step' that circumvented this difficulty. He proposed that globular clusters are distributed randomly around the galactic centre. We can see these clusters well away from the obscuring dust, and if we find the centre of their distribution in space it will pinpoint the invisible centre of the Galaxy.

In the early 1980s, Carlos Frenk and Simon White, at the University of California at Berkeley, refined Shapley's method, using the most recent measurements of globular clusters. They found that the galactic centre is about 22 000 light years away.

Another pioneer of the Milky Way was Walter Baade, who studied variable stars near the galactic centre, visible through gaps in the dust. The most prominent gap, in Sagittarius, is now known as 'Baade's Window'. In 1983, astronomers used the Infrared Astronomical Satellite to observe the dust-piercing infrared from many other variable stars near the galactic centre. Put together, the results on variable stars suggest that the Galaxy's centre lies at a distance of 26 000 light years.

Researchers have also used the scale model provided by the motions of remote star clusters and gas clouds. If we can pin down the actual distance to any of these, then we can derive the scale of the model, and so deduce the distance to the Galaxy's centre indirectly. No individual distances are known well enough to provide a definitive answer, but an average of many roughly known distances in the scale model tells us that we lie about 25 000 light years from the Galaxy's centre.

Finally, we can turn to radio telescopes that can see through the dust clouds and reveal the gas clouds in the Galaxy's core. In the late 1980s, the American radio astronomer Mark Reid studied carefully the motions of young stars in the gas cloud Sagittarius B2 that lies only a few hundred light years from the galactic centre. These stars energize the gas nearby, so they emit powerfully as natural masers – the microwave equivalent of lasers. The Doppler shift in the wavelengths from these masers revealed the natural spread in the speed of these stars: by comparing this velocity with the apparent motion of the masers across the sky, Reid could work out the distance to Sagittarius B2. Considering the limitations in the accuracy of the method, this is the same as the distance to the Galaxy's centre, and it came out as 23 000 light years.

Some of these methods are more accurate than others. Recently, both Reid and Michael Feast, of the South African Astronomical Observatory, have independently worked out the average of all the results, giving each a weight appropriate to its likely accuracy. They both conclude that the galactic centre is 25 000 light years away, with a possible error of 2000 light years.

kilometres per second. There were several other features – fragments of arms, arcs or jets – that seemed to be sharing in a general exodus from the galactic centre. Always a man ahead of his time, Oort suggested that there could have been some kind of outburst there.

At the time, he got little support. In the late 1950s, 'active galaxies' – those with energetic or disturbed nuclei – had yet to be recognized. The discovery of quasars lay years in the future, and the whole notion of cosmic violence held little sway. Astronomy was, in the main, a slow, conservative pursuit – not the subject at the cutting edge of science that it is today. And so Oort's contemporaries preferred either to believe that

FACING PAGE Unlike William Herschel's 'tunnels in space' (which turned out to be dark, obscured regions), here is a region that is just the opposite: Baade's Window. The reddened stars next to the highly obscured area at top right are very distant stars, close to the galactic centre, that are visible through a gap in the dust. Walter Baade discovered several RR Lyrae variable stars in the region and, by looking at their luminosity and distribution, made the first measurements of the distance to the galactic centre. His work has been recently updated by measurements at radio and infrared wavelengths, which all place the galactic centre at a distance of about 25 000 light years.

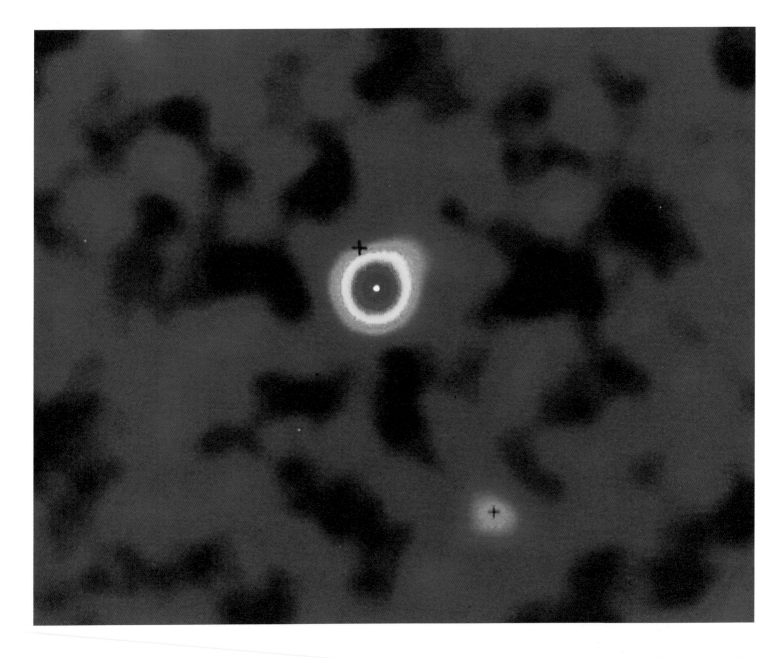

The bright 'galactic centre' gamma-ray source (colour-coded image at centre) and the galactic centre itself (marked by a cross) do not quite coincide. Until recently, this was thought to be caused by uncertainty in the gamma-ray measurements. Now, however, the Russian Granat spacecraft has pinned down the source of gamma rays to an object, the Great Annihilator, over 300 light years from the galactic centre – which means that the radiation is not evidence for a massive black hole, as once thought.

the errant arms near the galactic centre were indulging in some kind of 'resonance' activity (swinging in and out), or that the distribution of matter there was bar-shaped. Barred spiral galaxies have elongated nuclei instead of round ones, and the continually changing orientation of a central bar can mimic outward motions. There has been more recent evidence that our Galaxy has at least a 'mini-bar' at its centre, but this cannot explain all the activity at the heart of the Milky Way.

As radio telescopes continued to improve throughout the 1960s and 70s, enabling astronomers to see finer and finer detail, it became obvious that the area around the galactic centre was like no other in the Galaxy. It emerged that the central 1500 light years contains a huge reservoir of gas. The gas – which is more concentrated than anywhere else in the Milky Way – is enough to make 100 million stars.

With the discovery of more expanding features inside the gas reservoir, and a growing realization that the centres of a number of other galaxies were disturbed and disrupted, astronomers rose to the challenge and began to tackle the galactic centre head-on. Over the past quarter of a century, it has been attacked with a battery of instruments tuned to every wavelength that will penetrate the intervening murk – first radio waves, but now infrared, X-rays, and even gamma rays. We now have such an incredibly detailed picture of the centre of our Galaxy that we can see features there comparable in size with the diameter of Saturn's orbit about the Sun. And although there's still no complete agreement as to what is going on, the lure of the galactic centre is irresistible – after all, it is the closest example of a galaxy's nucleus that there is in the Universe.

The picture we have today of the galactic centre is of a region rather like the cosmic equivalent of Texas: a place where the biggest, richest, heaviest and most extreme objects in our Galaxy congregate. It is also characterized by several expanding features that, as Oort suggested so many years ago, point to some kind of explosive activity.

One of these is to be found just inside the gas reservoir, at a radius of about 500 light years from the galactic centre. Here, the density rises sharply and the gas clumps into several enormous clouds. These clouds form a ring, tilted at 20 degrees to the galactic plane, and expanding at 150 kilometres per second. Each cool, dense cloud is extraordinarily rich in molecules. As well as common varieties like carbon monoxide

(CO) and hydroxyl (OH), the clouds contain a chemist's cornucopia of exotic molecular species. Formaldehyde (H_2CO) and formamide ($HCONH_2$) rub shoulders with acetaldehyde (CH_3CHO) and ethanol (CH_3CH_2OH) – the latter perhaps demonstrating that there must be a 'bar' of sorts at the galactic centre! Writing about the molecular ring in a recent summary, Mark Morris – an expert on interstellar gas – says: 'It is difficult to escape the conclusion that this expanding molecular ring was set in motion by a very powerful explosion occurring at the nucleus about a million years ago'.

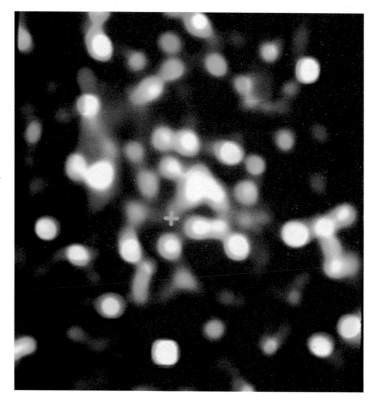

This colour-coded view of the innermost two light years of our Galaxy was made in near-infrared wavelengths. It shows a great many compact and extended infrared sources in the neighbourhood of the galactic centre, in particular IRS 16 (the source with the 'Mickey Mouse ears' at centre) and regions of warm dust. However, there appears to be no infrared counterpart of Sagittarius A*, the radio source (marked by a cross) that seems to mark the Galaxy's exact heart.

This wide-angle view of the galactic centre at infrared wavelengths takes in the scatter of nearby stars as well as the concentration of stars towards the galactic centre itself. Even though infrared wavelengths penetrate a good deal of the dust in the galactic plane, there is still some obscuration, as this image reveals.

The most massive molecular cloud in the Galaxy is located just inside the inner edge of the molecular ring, about 400 light years from the galactic centre. Sagittarius B2 (one of the original radio sources detected at the centre of the Galaxy) weighs in at a million solar masses, and is also the densest molecular cloud discovered so far. Virtually every species of interstellar molecule yet identified is present in this cloud, which, in addition, is home to several bright nebulae (all bigger than the Orion Nebula), plus a number of powerful masers.

These are all signs that Sagittarius B2 is a hotbed of star formation.

Closer in to the galactic centre itself, there are several smaller molecular clouds. Unlike the cold, ponderous clouds in the outer parts of the Galaxy, these are warmer and much more turbulent. The gas inside these clouds is at a temperature of −220 to −200 degrees Celsius (50 to 70 degrees above absolute zero, compared with normal molecular clouds at just 10 degrees above absolute zero), and moves around at speeds of

The millions of stars making up the galactic centre star cluster (shown here in an infrared view) are packed more densely than anywhere else in the Galaxy. In this image, nearby stars – visible through conventional telescopes – appear white or blue, while stars obscured by dust look extremely red.

20 to 30 kilometres per second – ten times faster than normal. Star formation could fuel this activity, but even frenetic star-birth is unlikely to be able to supply all the energy. Some astronomers are looking to another source – magnetism.

The first hints that strong magnetic fields have a major role to play in the galactic centre came in 1984. Using the radio telescopes of the Very Large Array (VLA) in New Mexico, Farhad Yusef-Zadeh made detailed observations of a feature called The Arc – a curved ribbon of gas about 100 light years away from the galactic centre. Yusef-Zadeh discovered that it looked like no other feature in the Galaxy. Instead of being an amorphous mass like most other nebulae, The Arc consists of numerous straight, parallel filaments of gas stretching across 150 light years of space. Each filament, however, is only half a light year wide.

The only force that could throw up such bizarre structures is magnetism – on a colossal scale. The filaments in The Arc are gigantic versions of the magnetic flux tubes called 'prominences' that we find on the Sun. Their size and linearity call for a magnetic field with a strength of about a milligauss – almost a thousand times stronger than the magnetic field we find elsewhere in the Galaxy. Because the filaments lie at right angles to the galactic plane, the magnetic field at the centre of the Galaxy must be 'poloidal' (north–south, like that of the

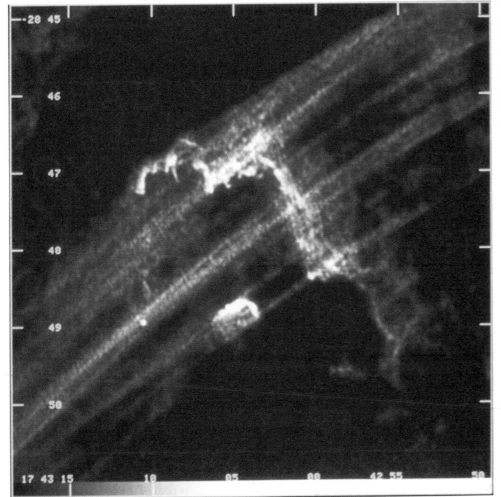

This radio image reveals the delicate structure of the narrow filaments making up The Arc. Each filament is less than half a light year wide, and the whole complex appears to be controlled by strong and extensive magnetic fields. The wisps of gas crossing the filaments make up a region nicknamed 'The Sickle'. It seems that the Sickle and the filaments interact with each other, and that this has led to the gas becoming ionized.

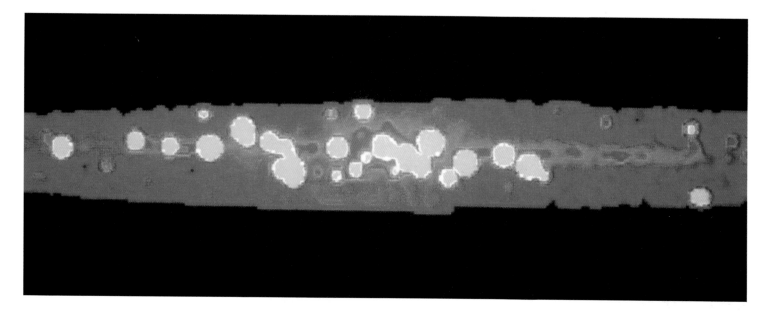

Exosat X-ray image looking towards the centre of the Galaxy, showing both discrete sources (blobs) and extended emission (from hot gas). The discrete sources probably have no connection with the galactic centre, and most are likely to be neutron stars in close binary systems lying along the line of sight. The hot gas, however, does appear to concentrate towards the centre of the Galaxy. It is at a temperature of 100 million degrees Celsius.

Earth). What causes the field is, as yet, unknown. But in the inner 100 light years of our Galaxy, the magnetic forces on clouds of gas must be comparable to the force of gravity that they feel – which may mean that much of the activity at the galactic centre may actually be driven by magnetism.

A whole menagerie of strange magnetic shapes has appeared on the images from the Very Large Array. Some, like the Mouse, may be caused by neutron stars tracking through the galactic centre region. The Snake, however, is apparently a piece of twisted magnetism. And the Sickle is a giant curve of hot gas linked up to the magnetic Arc.

The Arc may be part of an even larger structure. It sits at the bottom corner of an enormous feature shaped like a capital omega that extends 600 light years above the galactic plane. This 'galactic centre lobe' is 800 light years in diameter and looks like a huge magnetic loop. The lobe is nearly symmetrical about the galactic centre, which is why some astronomers interpret it as the surface of a 'chimney' of gas created by a rising blast from a central explosion.

Rising from the lobe is a vast curved jet, shaped like a banana and extending 13 000 light years into space. At first sight, it resembles the great jets found in some radio galaxies and quasars, but our Galaxy's example is less than one per cent as luminous as other galactic jets.

Inside The Arc, we home in on the galactic centre proper. The innermost 100 light years of our Galaxy contains a number of unique objects whose interpretation is still controversial. At the heart of the argument is whether or not the centre of our Galaxy harbours a massive black hole – and the extent to which it contributes to the activity there.

Surrounding the centre of our Galaxy is what appears to be a 'reservoir' of cool, un-ionized gas. Some of it is in molecular form, like the giant cloud complexes further out – but unlike the gas in the outer molecular ring, or in Sagittarius B2, there is no star-formation going on. Yet the gas here must be very dense to have avoided being torn to shreds by tidal forces. Possibly, the activity at the galactic centre is so disruptive that it prevents star-birth taking place.

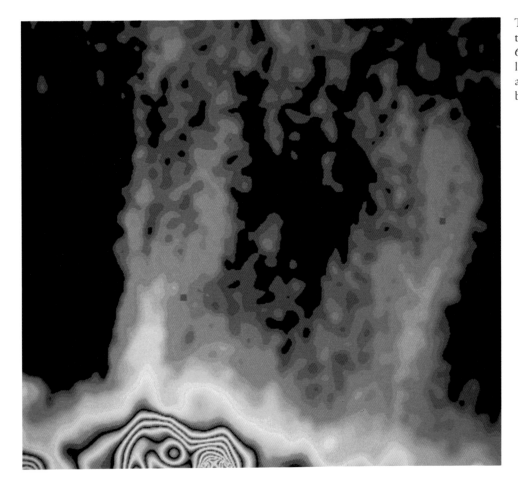

The galactic centre radio lobe, imaged here by the Nobeyama radio telescope in Japan, extends 600 light years above the galactic plane. Shaped like a huge magnetic loop, it is 800 light years across and straddles the galactic centre. It could be a 'chimney' of gas ejected by a central explosion.

Between 26 and 10 light years from the galactic centre, the gas orders itself into the 'circumnuclear disc'. Inside is a cavity that is relatively devoid of material, but which contains the object that holds the key to understanding the centre of our Galaxy – Sagittarius A.

One of the first radio sources to be detected near the galactic centre, Sagittarius A turns out to be two separate objects – Sagittarius A East and Sagittarius A West. Sagittarius A East is a bubble of hot gas that is probably a supernova remnant lying just behind the galactic centre (a similar remnant has just been discovered south-east of the centre). Sagittarius A West is a warm but un-ionized source that appears to sit bang in the middle of the Galaxy. At its heart is a mysterious compact radio source called Sagittarius A*. Many astronomers claim that this is the black hole that drives the activity at our galactic centre.

Although support for a black hole at the centre of the Galaxy has been steadily growing over the past few years, things are far from cut and dried. The 'black hole lobby' points to the fact that there has definitely been some kind of explosive activity centred on the Galaxy's core. There is also evidence that the galactic centre contains a considerable concentration of mass, and that there is *something* at the centre responsible for generating large amounts of ionizing radiation. A black hole could be responsible for all these phenomena – but there are other possibilities.

You can estimate how much mass there is at the galactic centre by studying the dynamics of the circumnuclear disc. If

Emerging straight from the galactic centre (bottom centre) is a vast jet recently discovered by radio astronomers in Australia and Germany. At its base is the galactic centre radio lobe; from there, the jet extends some 13 000 light years into space. There is a small possibility that the jet is part of a supernova remnant lying along the line of sight, but the coincidence with other structures at the galactic centre makes this seem unlikely. If it is really a jet, it is nevertheless 100 to 1000 times less luminous than its counterparts in other galaxies.

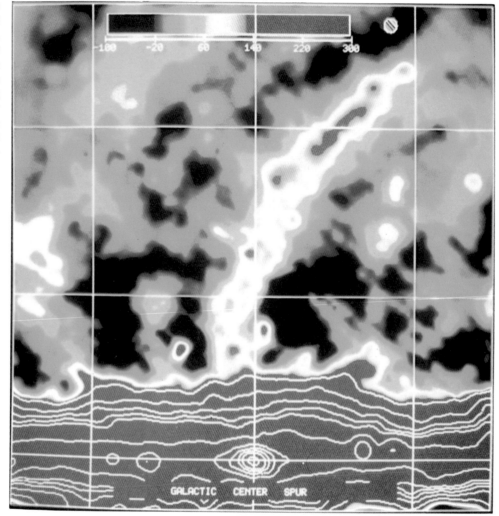

it is in a stable orbit about the centre of the Galaxy, its speed of rotation will tell you how much mass there is inside. The dynamics suggest that the mass is split into two 'regimes': a distributed mass (stars) and a point mass (the black hole). Currently, the best measurements give a mass of about five million solar masses in stars, and three million solar masses in a black hole.

Measurements made at the edge of the 'cavity' surrounding Sagittarius A West pin down the mass distribution closer in to the centre. Here, the gas forms an ionized ring – the 'western arc' – that goes halfway around the cavity. If this gas

is in a straightforward orbit, then its speed of 110 kilometres per second leads to a total mass of about four million solar masses – two million in stars, and two million in the black hole. Black hole supporters believe that the very fact that the cavity exists is evidence that the region has been 'scoured' by violent winds from the accretion disc that surrounds the hole itself.

But the cavity is not completely empty. Quite apart from a clutch of central objects (including Sagittarius A West and several infrared sources), the region is criss-crossed with streamers of ionized gas. The western arc, together with the northern and eastern arms, make up a kind of 'mini-spiral' at the galactic centre. The velocities of this gas are very difficult to disentangle. While the outer parts of the 'spiral' seem to be in stable orbits, the gas closest to the galactic centre has speeds of up to 400 kilometres per second. Opinion is currently divided as to whether these high velocities are caused by orbital motions, outflows of gas, or even inflows of gas.

What is undisputed is that the gas is ionized. Because there are no stars embedded there, what is doing the heating? At first, astronomers assumed the culprit was Sagittarius A West, and Sagittarius A* in particular. But new research by Farhad Yusef-Zadeh and Fulvio Melia with the VLA telescope is casting doubt on this interpretation. In 1990, Yusef-Zadeh used the VLA to make observations of the infrared source IRS 7. This is probably a red supergiant star only one light year from the galactic centre. Yusef-Zadeh discovered that it has a comet-like plume pointing away from it, as if the star were pushing through a dense medium, or if some external

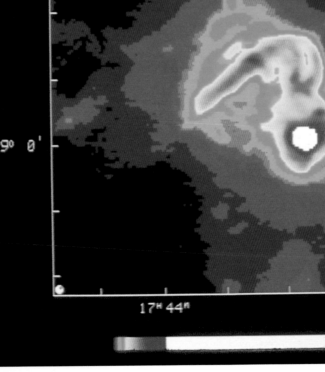

This colour-coded radio image of the central one degree of our Galaxy (an area roughly 450 light years across) covers the region of the sources in Sagittarius. At top left is the giant molecular cloud Sagittarius B2, while Sagittarius A, which surrounds the exact centre of the Galaxy, lies to the bottom right. Between the two is the mysterious 'Arc' feature, which is orientated at right angles to the galactic plane.

FACING PAGE The innermost 200 light years of the Galaxy, imaged by the radio telescopes of the Very Large Array, shows detailed structure in both The Arc and in Sagittarius A. Most striking are the narrow filaments in The Arc, which are thought to be the result of strong magnetic fields at the centre of the Galaxy.

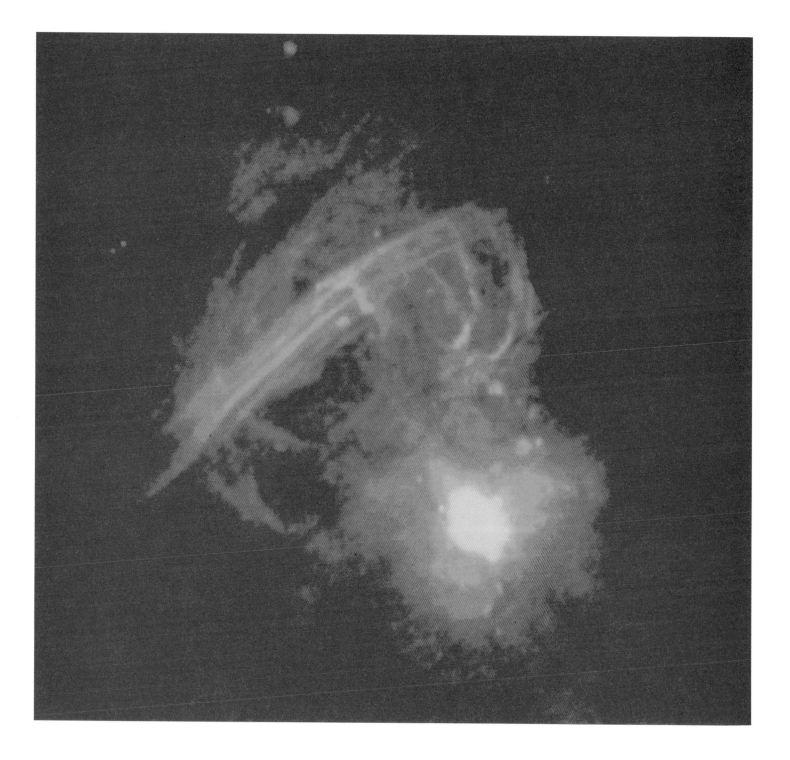

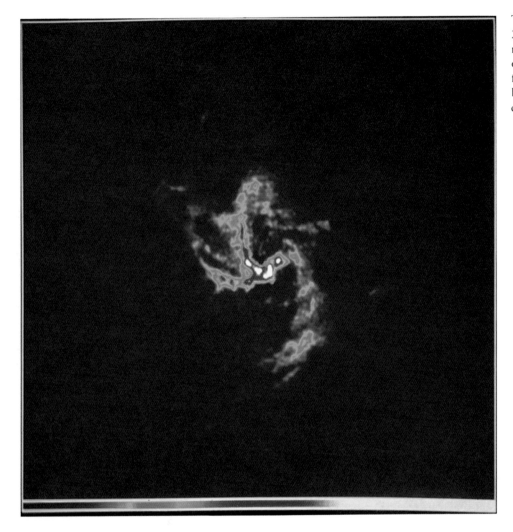

This view of the ionized gas in the innermost 20 light years of our Galaxy, made by the VLA, reveals the 'mini-spiral' that some researchers once believed were precessing jets emerging from an explosion at the galactic centre. Now, better measurements suggest that the 'ovally' distributed gas is part of a ring, while the 'bars' are infalling material that is being heated up.

wind were blowing material away – but where is the wind coming from?

Yusef-Zadeh and Melia looked at the direction of the axis of IRS 7's tail to discover if they could trace the wind's origin. The axis passed close to Sagittarius A* – but not close enough. However, just 0.15 light years from Sagittarius A*, and in exactly the right direction, is the infrared source IRS 16: a dense cluster of hot blue stars that generates a powerful stellar gale that gusts at 700 kilometres per second. IRS 16 is undoubtedly the source of the wind blowing on IRS 7. It is also a very active star cluster, with internal gas velocities of up to 1000 kilometres per second. Several infrared astronomers have proposed that the young blue stars of IRS 16 – and not mysterious Sagittarius A* – are responsible not only for creating an interstellar gale, but also for the ionizing radiation that bathes the galactic centre.

Another pillar of support for the galactic-centre black hole has also crumbled recently. One of the most powerful arguments in favour of a black hole was the discovery in 1977 of gamma radiation coming from the direction of the galactic centre. The energy of the radiation appeared to be the clincher: it was the 511 keV line from the annihilation of positronium.

Sagittarius A East and West both appear on this image of the central 100 light years of the Galaxy. Made by the VLA, the image is coded so that thermal sources (hot gas) appear red, while those emitting synchrotron radiation are blue. The 'mini-spiral' of Sagittarius A West is prominent in thermal radiation; to the left is synchrotron radiation coming from Sagittarius A East. This 'bubble' of ionized gas is probably a supernova remnant.

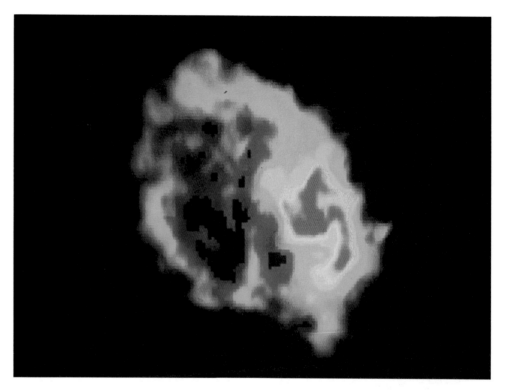

This unholy alliance of an electron and its antimatter equivalent, a positron, can only be created in very high energy environments. And although measurements subsequent to 1977 were sometimes uncertain about the exact strength of the line, these were interpreted as meaning that the source was rapidly variable – an indicator of its small size.

The problem was that the position of the gamma-ray source was not accurately known – early measurements were sometimes several degrees out. But on 13 October 1990, the Russian Granat spacecraft finally pinned it down. It turned out to be 45 arcminutes away from the galactic centre – 340 light years from the hub of the action. It coincides with an X-ray source called 1E1740.7−2942 that had been discovered ten years before by the Einstein satellite. In 1992, observations with the VLA showed that this 'Great Annihilator' is shooting two beams of positrons in opposite directions: the beams travel several light years before the positrons are destroyed by electrons. The beams may well come from the accretion disc surrounding a black hole that weighs in at 8 to 15 solar masses

– but it isn't in the same league as the massive galactic centre beast that astronomers set out to find.

However, we dismiss Sagittarius A* at our peril. For a start, it really does appear to be situated at the dynamical centre of the Galaxy. If it were not, it would have a high orbital speed (like Mercury circling the Sun) – but its velocity seems to be less than 40 kilometres per second. If it does mark the centre of the Galaxy, then it must be a unique object.

But what *kind* of object is Sagittarius A*? The problem in finding out lies in its incredibly small size: it subtends an angle of just two-thousandths of an arcsecond, which at the distance of the centre of the Galaxy corresponds to 13 astronomical units (13 times the distance from the Sun to the Earth). Recently – using the techniques of Very Long Baseline Interferometry (VLBI) – it has been possible (at least partly) to resolve the source. It appears to be elongated, with one axis twice as long as the other – just as you would expect if you were viewing a tilted disc. It is also very bright for its small size.

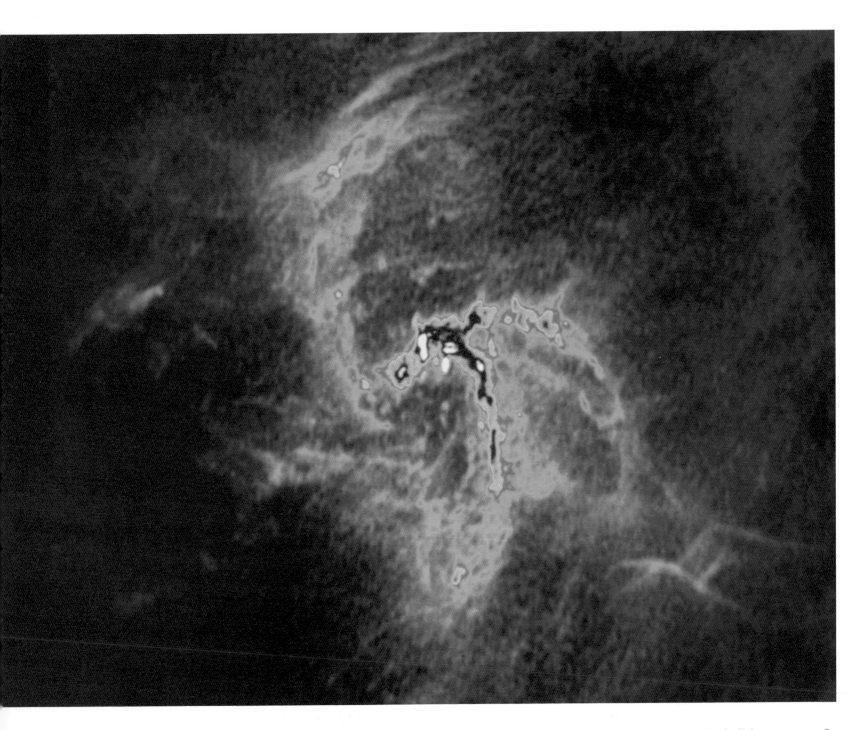

This extremely high-resolution image of Sagittarius A West – made with the VLA – shows details smaller than one-tenth of a light year across. In particular, it reveals the central radio source, Sagittarius A*, as a white dot just above the central 'bar'.

The galactic centre includes a number of strange and unusual objects that do not appear to have counterparts elsewhere. Among them is 'The Mouse' – a comet-shaped radio source with a long tail stretching behind it. The Mouse lives in a region that appears to contain a number of supernova remnants. Under detailed examination, the Mouse seems to be moving through the interstellar medium, leaving a wake behind it. If the source is at the distance of the galactic centre, than its tail is 100 light years long. It is possible that the Mouse is a runaway neutron star, with the 'tail' composed of ejected material; alternatively, it may be matter from the surrounding medium accelerated by the neutron star's strong magnetic field.

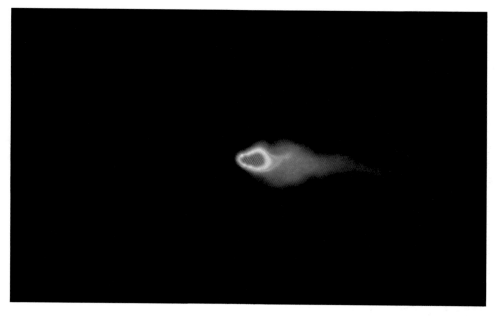

BOX 2. Quasars

Since the 'wavelength revolution' in astronomy began, following the end of the Second World War, astronomers have turned up a bewildering variety of galaxies with disproportionately bright or active nuclei. Although they make up only one per cent of all known galaxies, their output is so enormous that they have produced nearly as much energy as all the others put together.

First to be discovered were Seyfert galaxies – apparently normal spiral galaxies whose nuclei were dominated by light coming from a tiny, starlike point. Surrounding the nucleus, the astronomer Carl Seyfert discovered hot, dense clouds moving at speeds of up to 30 000 kilometres per second, evidentally churned up by the energy of the brilliant nucleus. Because the brightness of the nucleus varied rapidly, sometimes on timescales of months, it had to be very small – less than a light year across. But the luminosity of this nucleus was phenomenal – equivalent to the luminosity of a whole galaxy of 100 billion stars. The problems lay in understanding how such concentrations of energy were possible, and in how the 'central engine' worked.

Radio astronomers then found a class of active galaxy that gave out as much energy in radio waves as in light. The radio waves were beamed in narrow jets of electrons from the centre of the galaxy – usually a giant elliptical – and, on escaping the galaxy itself, the jets rammed into the intergalactic medium to produce enormous clouds.

Following the discovery of radio galaxies, astronomers found radio waves coming from apparently normal stars. But these stars' spectra were like nothing ever seen before – until a Dutch–American astronomer, Maarten Schmidt, realized that the spectra could be understood if the 'stars' lay further away than nearly all the known galaxies. Later, it transpired that most 'quasars' (the contraction of 'quasi-stellar radio source') do not emit radio waves. But they all shared in the distinction of being the most remote objects in the Universe, and in looking like the centres of Seyfert galaxies minus the surrounding galaxy. The nuclei of quasars, like those of Seyfert galaxies, were also variable – allowing an estimate of their size (and hence their energy) to be made.

The power of a typical quasar turned out to be even more

daunting than that of a Seyfert or a radio galaxy. What was more, these remote objects were emitting radiation right across the electromagnetic spectrum. Some of their cousins – the BL Lac Objects, or 'blazars' – compounded the energy problem by changing in brightness by a factor of up to 100 in just a few months. Astronomers, however, were confident that what they were seeing in all these cases were scaled-up versions of Seyfert galaxies. In fact, a few quasars were just close enough for the surrounding structure of the galaxy to show up dimly.

Until recently, it wasn't at all clear how all these manifestations of activity were related. But now astronomers believe that they are all aspects of the same phenomenon: the Active Galactic Nucleus, or AGN. The key to the activity is the gravity of a massive black hole, millions or billions of times the mass of the Sun, lying right at the centre of the galaxy. Surrounding it is an accretion disc: a glaring whirlpool of gases from disrupted stars pouring into the hole. Further out is a fat doughnut of cooler dust and gas, called the torus. If the AGN is beaming radio jets, they stream out from the inner edges of the accretion disc along its axes in opposite directions (although it is still unclear exactly how they are generated). The AGN, then, looks like a doughnut with a straw stuck through it – and in place of the jam, it has a diamond-bright accretion disc surrounding a pinprick of dark that is the supermassive black hole.

Whether we see a Seyfert, a radio galaxy, a quasar or a blazar depends on the angle at which we are viewing the AGN, and the activity of the AGN itself. A view straight down the jet and into the central 'engine' gives you a blazar – the light intensity varies according to the rate the black hole gobbles up the gas in the accretion disc. 'Peeping over the top' of the torus and *just* seeing the core will result in a Seyfert (for weaker AGNs), or a quasar (for powerful AGNs). A side-on view, in which the accretion disc is hidden, gives a radio galaxy.

Deciding whether the Milky Way has an AGN at its heart is difficult – partly because of the obscuration towards the galactic centre, and also because galactic nuclei generally become less active as time goes by. There are many quasars at great distances in the Universe, but none nearby; in other words, quasars were common when we look back into the past, but they are not now. The other question to answer is whether only one per cent of galaxies undergo quasar-like activity – or if *all* galaxies spend one per cent of their lives being active. If the latter is the case, then we should certainly expect to find the 'equipment' for an AGN in place at the centre of our Galaxy.

These findings have convinced many astronomers that Sagittarius A* is the accretion disc surrounding the black hole at the centre of our Galaxy. But one thing is certain: it is currently resting. If it were actively accreting material, we would see the disc glaring with infrared radiation. And the funnelling-in of stars and gas would leave tell-tale signs in the star distribution near the galactic centre that we certainly don't see.

There are ways in which we can pin down the nature of Sagittarius A* with more certainty. VLBI from space will give us a much better picture of the source itself. And if there is a black hole inside Sagittarius A*, it may manifest itself in other ways. The strong gravity of black holes warps space in their vicinity, causing 'gravitational lensing' of objects lying directly behind. Yusef-Zadeh is proposing to use the Very Long Baseline Array – a radio telescope that will effectively stretch from Hawaii to the Caribbean – to look for such effects. And new,

sensitive array detectors being developed for infrared astronomy will be able to map the star distribution around Sagittarius A* much more accurately.

In the end, the question boils down not to: *is* there a black hole at the centre of the Galaxy? It is more: *how* much has it shaped events there? Some astronomers readily acknowledge that there may be a small black hole in residence, but that the 'explosive' events in our Galaxy's history are driven by episodes of rapid star-formation, as in starburst galaxies. On this basis, our Galaxy may undergo another starburst in about 10 million years time, when the innermost molecular clouds lose energy by colliding with one another and spiral in to the galactic centre.

But most astronomers see the need for a black hole at the Galaxy's heart. The conditions there are so strange, so unlike anywhere else in the Galaxy that they seem to demand some

BOX 3. Starburst galaxies

Recently, a completely new class of active galaxies has come to light, whose energy is powered not by a central black hole, but by a violent burst of star-formation. One of the prototype 'starburst' galaxies – M82 – was once thought to have a quasar-like nucleus. The highly-disturbed, fast-moving filaments surrounding the galaxy were thought to be the result of a massive explosion, and the dense obscuration of the nucleus by dust was believed to stem from the same cause. But M82's nucleus was naggingly un-quasar-like in some respects – particularly in that it was lacking a compact radio source. However, it was emitting strongly in the infrared – more than ten times as powerfully as our Galaxy – and so there was clearly a prolific source of energy there.

The pieces began to fall into place when astronomers realized that the filaments – instead of being ejected from the galaxy – were actually streaming onto it. Also, the group in which M82 lives is rich in intergalactic gas driven out of the member-galaxies by interactions. M82 is running into this gas, which is being channelled towards the nuclear regions to form stars. It therefore owes its luminosity not to quasar-type explosions, but to a sudden, intense outbreak of star-formation. The starburst spawns many massive stars, which produce an immense amount of radiation and powerful stellar winds. In addition, these stars live for only a short time, and emit even more energy when they die as supernovae.

Other nearby galaxies – many in small groups rich in intergalactic gas – are now known to be starburst galaxies. The edge-on spiral NGC 1253 is one. So too is NGC 1068, which has both a compact nucleus like a Seyfert galaxy *and* a ring 10 000 light years out from the centre in which a starburst is taking place. But most of the tens of thousands of starburst galaxies now known are distant ones discovered by the IRAS infrared satellite. Some of these are more than 100 times more luminous in the infrared as our Galaxy – as brilliant as the most powerful quasars. Most (but not all) appear to be interacting or merging with another galaxy, which provides a ready source of gas for the starburst.

Could the Milky Way have a starburst nucleus? A number of astronomers believe that it has, and point out that some

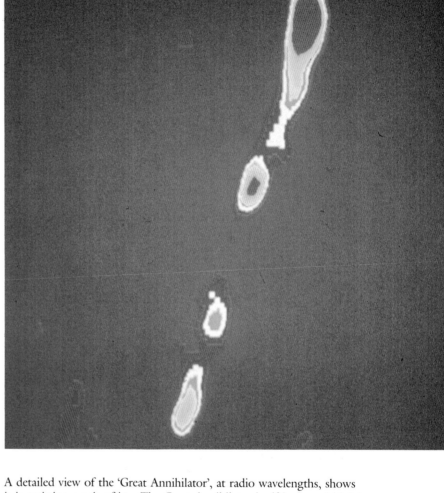

A detailed view of the 'Great Annihilator', at radio wavelengths, shows it is emitting a pair of jets. The Great Annihilator itself is the red blob in the middle of this image, and probably consists of a black hole tearing matter from a companion star. In the process, it generates two beams of positrons. These travel two or three light years before they hit ordinary electrons and annihilate in a burst of energy, giving bright spots at the end of the jets here and emitting gamma rays copiously.

of the dense molecular clouds close to the centre may spiral in to fuel a burst of star-formation in 10 million years or so. And it is also possible that interactions with the Magellanic Clouds could fuel a starburst, but on longer timescales.

kind of exotic denizen to drive events. Shklovskii was absolutely right: there is an 'outstanding peculiarity' at the galactic centre. By continuing to crack down on our own galactic nucleus – the closest we can study by a long shot – we will arrive at a better understanding as to what makes our own, and other, galaxies tick.

A galaxy that gives out a lot more heat than light, M82 is the prototype 'starburst galaxy'. Its central regions are undergoing a giant spasm of star-birth, generating an immense amount of infrared radiation. Light from the newborn stars is pouring out of the galaxy and illuminating nearby gas, visible in bright red in this colour-enhanced photograph.

FURTHER READING

General

Astronomical Objects for Southern Telescopes, by E. J. Hartung, CUP, 1968.

The Astronomy and Astrophysics Encyclopedia, edited by Stephen P. Maran, CUP 1992.

Atlas of Deep-Sky Splendors, by Hans Vehrenberg, Sky Publishing, 1983.

The Brightest Stars, by Cornelis de Jager, D. Reidel, 1980.

The Cambridge Atlas of Astronomy, edited by Jean Audouze and Guy Israël, CUP, 1985.

Catalogue of the Universe, by Paul Murdin, David Allen and David Malin, CUP, 1979.

Celestial Objects for Common Telescopes, by the Rev. T. W. Webb, Dover, 1962.

Classics in Radio Astronomy, by Woodruff T. Sullivan III, D. Reidel, 1982.

Collins Guide to Stars and Planets, by Ian Ridpath and Wil Tirion, Collins, 1984.

The Color Atlas of Galaxies, by James D. Wray, CUP, 1988.

Colour Star Atlas, by John Cox and Richard Monkhouse, George Philip, 1991.

Colours of the Stars, by David Malin and Paul Murdin, CUP, 1984.

The Constellations, by Lloyd Motz and Carol Nathanson, Aurum, 1988.

Cosmic Discovery, by Martin Harwit, Basic Books, 1981.

The Discovery of our Galaxy, by Charles A. Whitney, Angus and Robertson, 1971.

The Expanding Universe, by Robert Smith, CUP, 1982.

The Historical Supernovae, by David H. Clark and F. Richard Stephenson, Pergamon, 1977.

A History of Astronomy, by A. Pannekoek, Allen and Unwin, 1961.

Exploring the Southern Sky, by Svend Laustsen, Claus Madsen and Richard M. West, Springer-Verlag, 1987.

The Fullness of Space, by Gareth Wynn-Williams, CUP, 1992.

Galaxies, by Paul W. Hodge, Harvard UP, 1986.

Messier's Nebulae and Star Clusters, by Kenneth Glyn Jones, CUP, 1991.

The Milky Way, Bart J. Bok and Priscilla F. Bok. Harvard, 1981 (fifth edition).

The New Astronomy, by Nigel Henbest and Michael Marten, CUP, 1983.

Observing the Constellations, by John Sanford, Mitchell Beazley, 1989.

Sky Atlas 2000.0, by Wil Tirion, CUP, 1981.

Sky Catalogue 2000.0 (vols 1 & 2), edited by Alan Hirschfeld and Roger W. Sinnott, CUP, 1982 & 1985.

A Source Book in Astronomy, by Harlow Shapley and Helen E. Howarth, McGraw-Hill, 1929.

The Southern Sky, by David Reidy and Ken Wallace, Allen & Unwin, 1987,

The Space Atlas, by Heather Couper and Nigel Henbest, Dorling Kindersley, 1992.

Star Names, their Lore and Meaning, by Richard Hinckley Allen, Dover, 1963.

The Stars, by Heather Couper and Nigel Henbest, Pan, 1988.

The Universe from your Backyard, by David J. Eicher, CUP, 1988.

A View of the Universe, by David Malin, CUP, 1994.

Conference proceedings

The Center of the Galaxy, edited by Mark Morris, Kluwer, 1989.

The Galaxy, edited by Gerry Gilmore and Bob Carswell, D. Reidel, 1987.

The Galaxy and the Solar System, edited by Roman Smoluchowski, John N. Bahcall and Mildred S. Matthews, University of Arizona, 1986.

Kinematics, Dynamics and Structure of the Milky Way, edited by W. L. H. Shuter, D. Reidel, 1983.

The Large-Scale Characteristics of the Galaxy, edited by W. B. Burton, D. Reidel, 1979.

Light on Dark Matter, edited by F. P. Israel, D. Reidel, 1986.

The Milky Way Galaxy, edited by Hugo van Woerden, Ronald J. Allen and W. Burton Butler, D. Reidel, 1985.

Regions of Recent Star Formation, edited by R. S. Roger and P. E. Dewdney, D. Reidel, 1982.

The Spiral Structure of our Galaxy, edited by W. Becker and G. Contopoulos, D. Reidel, 1970.

Reviews and news items

Original research on the Galaxy is published principally in *Astronomical Journal*, *Astronomy & Astrophysics*, *Astrophysical Journal*, the *Monthly Notices* and *Quarterly Journal of the Royal Astronomical Society* and *Nature*.

Review articles occasionally appear in *Nature* and *Annual Reviews of Astronomy and Astrophysics*.

Popular articles and news items can be found in the magazines *Astronomy*, *Astronomy Now*, *Focus*, *New Scientist*, *Scientific American*, *Sky & Telescope* and *Sky & Space*.

PICTURE CREDITS

Many of the photographs and illustrations in the book are available from **Science Photo Library (SPL)**. Full addresses of all major sources are given at the end of the credits. Abbreviations: b = bottom, l = left, r = right, t = top.

Page 3.   Roger Ressmeyer, Starlight/SPL.
Page 4.   *The Origin of the Milky Way*, by Tintoretto. © National Gallery.
Page 5.   SPL.
Page 6.   Korean Planisphere from *ISIS* **XXXV**, part 4, no. 102. © University of Chicago Press.
Page 7.   James Barlow, map of the Southern Celestial Hemisphere. © British Museum.
Page 8.   Hencoup Enterprises/C. & M. Marten/SPL.
Page 9.   Royal Greenwich Observatory/SPL.
Page 11.   Royal Greenwich Observatory/SPL.
Page 12.   © Royal Astronomical Society.
Page 13t.   Hencoup Enterprises/C. & M. Marten/SPL.
Page 13b.   Harvard College Observatory/SPL.
Page 14.   Courtesy Prof. Owen Gingerich.
Page 16.   David Malin, © Anglo-Australian Telescope Board 1987.
Page 17.   Rev. Ronald Royer/SPL.
Page 18.   Roger Ressmeyer, Starlight/SPL.
Page 19l.   Hale Observatories. Courtesy AIP Niels Bohr Library.
Page 19r.   The Observatories of the Carnegie Institution of Washington.
Page 20.   Hencoup Enterprises/C. & M. Marten/SPL.
Page 21.   Hencoup Enterprises/C. & M. Marten/SPL.
Page 25.   Courtesy of Prof. Owen Gingerich.
Pages 26–7.   Courtesy Lund Observatory.
Pages 32–3.   Fred Espenak/SPL.

Page 34.   Astronomische Arbeitsgemeinschaft Bochum (AABO)/SPL.
Page 35.   Dr Eli Brinks/SPL.
Page 37.   Roger Ressmeyer, Starlight/SPL.
Page 38.   Space Telescope Science Institute/NASA/SPL.
Page 41.   Dr Jean Lorre/SPL.
Page 43l.   Dr Rudolph Schild/SPL.
Page 43r.   Dr Rudolph Schild/SPL.
Page 44.   David Malin, © Anglo-Australian Telescope Board 1981.
Page 47.   Royal Observatory, Edinburgh/AATB/SPL.
Page 48.   Robert H. McNaught/SPL.
Page 49.   Max-Planck-Institut für Radioastronomie/ SPL.
Page 50.   National Optical Astronomy Observatories/ SPL.
Page 51.   Royal Observatory, Edinburgh/SPL.
Page 58.   © European Southern Observatory.
Page 61.   NASA/SPL.
Page 63.   Prof. Carl Heiles/SPL.
Page 64.   Dr Rudolph Schild/SPL.
Page 65.   NASA/SPL.
Page 67.   SPL.
Page 68.   C. Powell, P. Fowler & D. Perkins/SPL.
Page 69.   Max-Planck-Institut für Extraterrestrische Physik/SPL.
Page 70.   © Infrared Processing and Analysis Center 1992, California Institute of Technology.

NAME INDEX

SUBJECT INDEX